Observer Design for Control and Fault Diagnosis of Boolean Networks

Zhihua Zhang

Observer Design for Control and Fault Diagnosis of Boolean Networks

Springer Vieweg

Zhihua Zhang
Kaiserslautern, Germany

Dissertation Technische Universität Kaiserslautern, 2021

ISBN 978-3-658-35928-7 ISBN 978-3-658-35929-4 (eBook)
https://doi.org/10.1007/978-3-658-35929-4

Responsible Editor: Stefanie Eggert
This Springer Vieweg imprint is published by the registered company Springer Fachmedien Wiesbaden GmbH part of Springer Nature.
The registered company address is: Abraham-Lincoln-Str. 46, 65189 Wiesbaden, Germany

Acknowledgement

Behind every Ph.D. thesis there are years of hard work. The work that resulted in this thesis could not have been accomplished without the support and encouragement of several persons that will be mentioned hereafter.

This thesis was written while the author was part of the team at the Institute of Automatic Control in Department of Electrical and Computer Engineering at the Technische Universität Kaiserslautern. First of all, I would like to express my deepest appreciation to PROF. DR. PING ZHANG, the head of the institute. Her perpetual encouragement, guidance and support enabled me to complete this work. Besides PROF. ZHANG, my sincere appreciation go to the other members of the evaluation board: PROF. DR.-ING. JAN LUNZE from Ruhr-University Bochum in Germany, and to the chairman PROF. DR.-ING. STEVEN LIU from Technische Universität Kaiserslautern in Germany.

I would like to warmly thank my college THOMAS LEIFELD for valuable discussions and helpful suggestions. It was my honor to work together with him in the framework of the project. Many thanks to MIRIAM STRAKE, DINA MIKHAYLENKO and RAPHAEL FRITZ for the proofreading of the manuscript and for hinting at a considerable number of flaws. Moreover, I want to thank my German teachers Mrs. Heilmann and Mr. Heilmann for the support during my time in Kaiserslautern.

Last but not least, I want to express my gratitude to my parents for their love, patience and encouragement.

Kaiserslautern
September 2021

Zhihua Zhang

Abstract

Systems biology is the computational modeling of biological systems (e.g. cellular processes), which enables scientists to study the interactions between entities within biological systems. Due to high complexity of cellular processes, a kind of parameter-free models called Boolean control networks (BCNs) can be used to approximate the qualitative behavior of biological systems. By using semi-tensor product of matrices, the dynamics of the BCNs are converted into a model similar to the standard discrete-time state-space model. This enables the solution of the control-theoretic problem of BCNs.

In control theory, a state observer can provide information on the internal states that can be used in many other applications, for instance, tracking control and observer-based fault diagnosis. As reconstructibility condition is necessary for the existence of a state observer, in this thesis explicit and recursive methods are developed for reconstructibility analysis. For state estimation, an approach to design Luenberger-like observer is proposed, which works in a two-step process (i.e. predict and update). It is proven that if a BCN is reconstructible, then an accurate state estimate can be provided by the observer no later than the minimal reconstructibility index. By considering different factors the approach is extended to enable the design of unknown input observer, distributed observers and reduced-order observer for a wide range of applications. The performance of the observers is evaluated thoroughly. Furthermore, methods for output tracking control and fault diagnosis of BCNs are developed. Finally, the developed schemes are tested with numerical examples.

Contents

About the Author

Zhihua Zhang was born on 09th September 1988 in Fujian, China, and lives currently in Kaiserslautern, Germany. He studied electrical/computer engineering at the Technische Universität Kaiserslautern under the supervision of Prof. Dr. Ping Zhang. His research interests include systems biology, optimal control, observer design and diagnosis for logical networks. Since June 2019, Zhihua Zhang works as an E&I engineer at the company BASF SE in Ludwigshafen, Germany.

Abbreviations and Acronyms

ASLS	Autonomous switched linear discrete-time system
BCN	Boolean control network
BN	Boolean network
DFA	Deterministic finite automata
RCP	Relational coarsest partition
STP	Semi-tensor product of matrices
UIO	Unknown input observer

General notations

$\neg$	negation
$\vee$	disjunction
$\wedge$	conjunction
$\mathcal{D}$	set {True, False} or {0, 1}
$\mathcal{D}^n$	n-dimensional Boolean vector $\underbrace{\mathcal{D} \times \mathcal{D} \times \cdots \times \mathcal{D}}_{n}$
δ_n^i	i-th column of the n-dimensional identity matrix I_n
Δ_n	set $\{\delta_n^i \mid 1 \leq i \leq n\}$
$\delta_n[i_1 \ \ i_2 \ \ \cdots \ \ i_k]$	logical matrix with $\delta_n^{i_j}$ as its j - th column
$\delta_n\{i_1 \ \ i_2 \ \ \cdots \ \ i_k\}$	sequence $\{\delta_n^{i_1}, \delta_n^{i_2}, \cdots, \delta_n^{i_k}\}$
$\mathcal{L}_{n \times t}$	set of logical matrices of dimension $n \times t$
$W_{[m,n]}$	swap matrix with index (m, n)
Φ_n	power-reducing matrix
$\mathbf{1}_n$	$[\underbrace{1 \ 1 \ \cdots \ 1}_{n}]^{\mathrm{T}}$
$\mathbf{0}_n$	$[\underbrace{0 \ 0 \ \cdots \ 0}_{n}]^{\mathrm{T}}$
$[v]_j$	the j-th entry of the vector v
$Col_i(M)$	the i-th column of the matrix M
$Col(M)$	set of columns of the matrix M
$Row_i(M)$	the i-th row of the matrix M
$Row(M)$	set of rows of the matrix M
$Blk_i(M)$	the i-th block of matrix M
$M_{i,j}$	the (i, j)-th element of the matrix M

$\mathrm{diag}(A_1, \cdots, A_n)$	a block matrix, in which the blocks on the main diagonal are the matrices $A_1, \cdots, A_n$ and the other blocks are zero matrices.
$M_{i,j}$	the (i, j)-th entry of the matrix M
$\odot$	Hadamard product
$\otimes$	Kronecker product
$\prod_{j=1}^{N} A_j$	$A_N A_{N-1} \cdots A_1$.
$\ltimes$	semi-tensor product
$\ltimes_{\mathcal{B}}$	Boolean product
$\|\cdot\|_\infty$	the infinity norm of a function
$\lvert\cdot\rvert$	absolute value or cardinality of a set
$\lceil\cdot\rceil$	ceiling function
$\mathbb{N}$	the set of nature numbers
$\mathbb{R}$	the set of real numbers
$\mathbb{Z}$	the set of all integers
$\mathbb{Z}_+$	the set of all positive integers
$n!$	the factorial of a positive integer n

Introduction

1

1.1 Motivation

Systems biology is the computational modeling of biological systems (e.g. cellular processes), which enables scientists to study the interactions between the entities within biological systems. Mathematical models built with differential equations give the relationship between physical quantities and the rate of change. In this way, they can represent the system dynamics in detail. However, in reality, almost all cellular processes are highly complex. Therefore, a large number of parameters in the differential equation models are required to be known, which is not feasible in practice. To estimate a large number of parameters of the system, regularization methods can be used (Zorzi and Sepulchre, 2016). Bornholdt (2008) pointed out that binary characteristics are shown in many molecular regulatory elements and one can observe bistable switches in metabolic and genetic networks, which are difficult to be considered in the differential equations. In contrast, Boolean networks (BNs) are well-studied parameter-free models, which were first introduced by Kauffman in Kauffman (1969) to model gene regulatory networks. The variables of BNs can take only two possible values (i.e. OFF "0" and ON "1") and each Boolean variable has a Boolean function assigned to it for updating. For a gene regulatory network if a Boolean variable has the value "0", then it means that following the Boolean rule a gene is inhibited. On the other hand, if a Boolean variable has the value "1", then this means that following the Boolean rule the corresponding gene is activated. BNs can be used to approximate the qualitative behavior of systems (Wynn et al., 2012) and can also be applied to describe discrete-event systems. Due to their advantages, BNs attracted increasing interest of many scientists. For instance, Akutsu et al., 2007 has studied the control problems of BNs. The result shows that control problems on BNs is $\mathcal{NP}$-hard in general.

Z. Zhang, *Observer Design for Control and Fault Diagnosis of Boolean Networks*,
https://doi.org/10.1007/978-3-658-35929-4_1

In reality, biological processes are influenced by external stimuli, for instance, temperature, drugs, concentration of oxygen. In addition, because of the limitation of measuring techniques, it is not possible to measure all the states directly. Besides, there exist unobserved post regulatory processes, such as protein-protein interactions. To take these factors into account, the input variables and output variables are included in the BNs, which results in Boolean control networks (BCNs). As pointed out by Akutsu et al. (2007), "one of the major goals of systems biology is to develop a control theory for complex biolgical systems". From this reason, it is meaningful to study control problems of BCNs. However, without an effective tool it is still a challenge. In recent years, a mathematical tool called semi-tensor product (STP) of matrices (Cheng et al., 2012) has been introduced by Cheng and co-workers. By using STP, the dynamics of the BCNs can be converted into a model that is quite similar to the standard discrete-time state-space model. This representation enables scientists to solve the control-theoretic problem of BCNs. In the dissertation, four different control-theoretic problems are considered, i.e. reconstructibility analysis, observer design, output tracking control and fault diagnosis.

In control theory, a state observer plays an important role, because it can provide information on the internal states that can be used in many other applications, for instance, state feedback controller and observer-based fault detection. Reconstructibility is the property of a system of which the final state can be determined uniquely with knowledge of the input and output trajectory. For BCNs, Fornasini and Valcher come to the conclusion in Fornasini and Valcher (2013) that the reconstructibility condition is necessary for the existence of a state observer. Hence, before observer design, it is necessary to study reconstructibility of BCNs. In the literature, there are several definitions for observability (Cheng and Qi, 2009; Zhao et al., 2010; Laschov et al., 2013; Fornasini and Valcher, 2013) and reconstructibility (Fornasini and Valcher, 2013; Xu and Hong, 2013; Zhang et al., 2016). In order to give a reason to study reconstructibility defined in Fornasini and Valcher, 2013 it is worth investigating the relationship between observability and reconstructibility of BCNs.

In order to design observers that can be used in wide range of applications, different factors shall be taken into account. In practice, disturbances, measurement noise and process noise often exist in practical systems. Quite often, disturbances cannot be directly measured or are too expensive to measure (Chen et al., 2016). An unknown input observer for BCNs shall be designed to provide a correct state estimate for BCNs with unknown inputs, such as disturbances and noises. Besides, in reality, almost all cellular processes are highly complex and large. However, to handle large-scale systems consisting of a huge number of states and measurements, high computational effort to execute centralized estimation is required. In recent years, the concept of distributed estimation has been introduced in Falcao et al.

(1995) for a faster and computationally more viable state estimation. Therefore, it is now more convenient to apply distributed estimation strategy, i.e. distributed observer design. Alternatively, some of the state variables can be accurately calculated with the knowledge of some output variables. For instance, as a special case, some state variables are directly measured. Hence it is not necessary to estimate the available state variables. By making use of this fact, it is meaningful to introduce an approach of reduced-order observer design to reduce computational complexity.

For the production of chemical and pharmaceutical products bioreactors are of growing importance. For instance, microalgae can be used as a renewable energy resource for the bio-fuel production. As mentioned in Abdollahi and Dubljevic (2012), one of the strategies to guarantee the efficiency of lipid production is to force the microalgae states to track a predefined reference trajectory. In practice, due to the limitation of measuring techniques, not all states can be directly measured. In addition, control design over infinite horizon is in some situations impractical due to the cost of the operation in a long-term treatment. Hence, it is meaningful to investigate the output tracking control problem of BCNs over a finite horizon. In some applications, frequent changes in control inputs are not allowed or not feasible. For instance, drugs can be viewed as a kind of control input in treatments. The pharmacological activity of some drugs may have a duration of more than a couple of hours in the organism and a change of drugs in a short time is therefore not advised. Furthermore, in gene regulatory networks, some undesirable states, transitions and inputs should be avoided during the treatment, because they may cause damaging effects, such as the deterioration of a disease (Xiao and Dougherty, 2007). Therefore, changes in control inputs and constraints shall also be considered in the controller design.

Fault diagnosis aims at detecting fault occurrence and isolating faults by means of observing the input and output trajectory generated by a system. A fault may cause an undesired deviation from the normal system's behavior. For instance, genetic alterations can cause diseases (Sridharan et al., 2012). If the derivation caused by the fault can not be tolerated by the system, then it is necessary to apply methods of fault diagnosis to detect and locate the fault. Hence, it is worth investigating model-based fault diagnosis of BCNs.

1.2 State of the Art

The following section will give an overview of the current state of research on BCNs, i.e. reconstructibility analysis, observer design, output tracking control and fault diagnosis as found in the literature.

Reconstructibility analysis

Reconstructibility is an important property of systems which indicates the convergence of state observers to the real state. Up to date, the problem of checking reconstructibility of BCNs has been considered in Fornasini and Valcher (2013), Zhang et al. (2016) and Xu and Hong (2013). The principle behind the approach proposed in Fornasini and Valcher (2013) is to find all distinct periodic state-input trajectories of the same minimal period and then to compare the corresponding output trajectories. If these output trajectories do not coincide, then the BCN is not reconstructible. However, when it comes to the application of this method, it is not easy to find all periodic state-input trajectories. Therefore, an approach is given in Zhang et al. (2016) which applies formal language and theories of finite automata. A pair of distinct states is considered as a vertex of a deterministic finite automata (DFA). The dynamic of the DFA is derived based on the system model. But as pointed out in Cheng et al. (2016), this approach involves an auxiliary machine, i.e. finite automata, so that the knowledge about formal language of finite automata is additionally required to understand their technique. In addition, recognizable languages of the DFA can only be verified for very small systems. Different from the approaches proposed in Fornasini and Valcher (2013) and Zhang et al. (2016), a matrix approach based on STP for reconstructibility analysis has been given in Xu and Hong (2013). However, to use the approach, the length of input and output trajectory to be considered should be known a priori. Besides, two high dimensional matrices need to be calculated for this method. Consequently, the matrix approach has high computational complexity. In computer science recursion is a commonly used method to simplify a complex problem into different sub-problems of the same type as the original problem. Hence, it is meaningful to introduce a recursive matrix approach for reconstructibility analysis by using STP, while reducing computational complexity.

Observer design

A state observer is a system that can provide an estimate of the internal state of a given system from the observed input and output trajectory. Recently, approaches to observer design for BCNs have been addressed in Fornasini and Valcher (2013), Xu and Hong (2013), and Fornasini and Valcher (2015a). In Fornasini and Valcher (2013), a Shift-Register observer and a Multiple States observer were introduced. However, a general mathematical formula for the system matrices of both observers has not been given so far. Furthermore, the Shift-Register observer does not provide any state estimate in the transient phase and requires very high computational effort due to its large system matrix. Though the Multiple States observer can provide state estimate all the time, a bank of observers needs to be used and it still has a high computational burden. The observer given in Fornasini and Valcher (2015a) is not described by means of a BCN, but by a two-step calculation procedure. Hence, it is very hard to study the observer performance. Different from that, the observer proposed in Xu and Hong (2013) is described in the form of a BCN. However, the relationship between convergence of these observers and reconstructibility have not yet been studied and the convergence of the state estimation has not been discussed. Besides, the existing approaches for observer design can not take unknown inputs, such as disturbances and noise, into account. Hence, it is required to consider unknown inputs in the model, such as disturbances and noises, and investigate the design of unknown input observer for BCNs with unknown inputs. As mentioned in Weiss and Margaliot (2019), state observers have exponential complexity in general. To handle large-scale systems consisting of a huge number of states and measurements, large memory consumption and computational effort are required to execute centralized estimation. For these reasons, the existing observers cannot deal with large-scale BCNs.

Tracking control

It is clear that if a state observer can provide correct state estimate of BCNs (i.e. real state), then based on the internal state many control problems can be solved, for instance, output tracking control. In the literature, the tracking of a constant output reference signal has been first studied in Fornasini and Valcher (2014a) by adopting the optimal control approach Fornasini and Valcher (2014b). Later, Li et al. (2015) proposed a state feedback controller design approach. However, tracking of a time-varying output reference trajectory has not been considered so far. In practical systems, the controller design may be subject to constraints. For finite horizon optimal control problem, Fornasini and Valcher (2014a) proposed an approach to consider constraints on states and transitions. However, until now input constraints were not considered.

Fault diagnosis

In the existing literature, the model-based fault detection problem of BCNs has been recently studied in Fornasini and Valcher (2015a), Fornasini and Valcher (2015b), and Sridharan et al. (2012). The approach proposed in Sridharan et al. (2012) applies a two-step principle. In the first step, an input sequence called a homing sequence is applied to drive the system to a known state. In the second step, a test signal will be generated to detect faults. However, the system is required to be reconstructible under the definition of reconstructibility given in Zhang et al. (2016). Besides, applying the homing sequence may lead to large time delays by fault detection. The approaches proposed in Fornasini and Valcher (2015a) and Fornasini and Valcher (2015b) apply a full-order observer to estimate the internal state. Based on the state estimate, a residual signal is generated for fault detection. However, using a full-order observer to provide the state estimate requires a high computational effort. To reduce computational complexity, the reduced-order observer can be applied, since some of the state variables can be accurately calculated with the knowledge of some output variables. The more state variables are measured, the lower computational effort is required. However, sometimes due to the limitation of measuring techniques and the cost of biological experiment, only a small amount of state variables can be directly measured. In this case, to further reduce computational complexity, an alternative reduced-order observer for fault detection will be proposed. After a fault has occurred, it is meaningful to study the fault isolation problem, which aims at localizing the system fault. Up to date, the fault isolation problem of BCNs has not yet been studied in the literature. Therefore, an approach based on a reduced-order observer for fault diagnosis will be proposed, which requires a lower computational effort than a bank of full-order observers for each fault candidate.

1.3 Organization

After this introduction, this dissertation will begin with the matrix expression of Boolean control networks (BCNs) in chapter 2. In this chapter, the definition and some preliminaries of the semi-tensor product (STP) of matrices used in this dissertation are given at first to support readers in understanding the proposed approaches. After that, it will be shown how to express BCNs into an algebraic form by using the STP.

In chapter 3, the reconstructibility analysis of BCNs is studied. At first, a comprehensive relationship between the various definitions of observability and reconstructibility of BCNs is given according to their definitions. The result shows that reconstructibility and observability are not equivalent in general. Then, an explicit method to check reconstructibility of BCNs by applying STP is introduced. The rationale behind the method is to directly build the mapping between input and output trajectory and final states. To facilitate the application to larger networks, a recursive method for reconstructibility analysis of BCNs is presented. The key is to show that the mapping for explicit method can be determined in an iterative procedure, which is stopped early for unreconstructible BCNs. In order to check reconstructibility of large-scale BCNs, a sufficient condition is given. The basic idea is to find the connections among network nodes, which do not play any role in state estimation. Cutting the connection, the large-scale BCNs becomes a system with acyclic structure. For BCNs with unknown inputs, an approach is introduced by analyzing the unknown input decouplability. To demonstrate the results, numerical examples are also given.

In chapter 4, the observer design of BCNs is considered. In addition, the performance of the proposed observer is studied. At first, the Luenberger-like observer design approach for BCNs is proposed to facilitate an online implementation and the relationship between the Shift-Register observer and the Luenberger-like observer is studied. The basic idea is to let the observer work in a two-step process (i.e. predict and update). In the prediction process, the state estimate will be predicted based on system equation. After that, the state estimate is updated with current observation information. If a BCN is reconstructible, it is proven that the state estimate provided by the Luenberger-like observer always contains the real state and converges to the real state at a time no later than the minimal reconstructibility index. After that, an approach is proposed for the design of unknown input observer (UIO). If the BCN is unknown input decouplable, then the UIO observer can provide the correct state estimate unaffected by unknown inputs. Based on the state estimate provided by the UIO observer, unknown inputs of the BCN can be estimated further. For a larger BCN, an approach for reduced-order observer design based on the reducible state variables is proposed. It is shown that, similar to the Luenberger-like observer, the state estimate provided by the reduced-order observer also converges to real state at a time no later than the minimal reconstructibility index. Additionally, the reduced-order observer requires lower computational effort than the Luenberger-like observer. In the case of a reconstructible large-scale BCN, an approach for distributed observer design is introduced to facilitate state estimation for large-scale BCNs. Numerical examples are given to illustrate the feasibility of the proposed approaches.

By making use of the information about the internal state provided by a state observer, an approach to design tracking controllers for BCNs to track a time varying reference output trajectory is proposed in chapter 5. To achieve this goal, necessary and sufficient conditions for the trackability of a time-varying reference output trajectory of finite length are given. The basic idea is to construct and analyze the indistinguishability set containing all the states that can generate the output at each time step. An approach to determine a control sequence for a trackable reference output trajectory is proposed, in which the control sequence is selected backwards based on the analysis of the indistinguishability set. If the reference output trajectory is not trackable, then two approaches are provided for the design of the optimal control sequence with the purpose of minimizing the tracking error. The key is to formulate the tracking problem as an ℓ_1 or ℓ_∞ optimization problem. To avoid frequent changes in control inputs, the proposed ℓ_1 and ℓ_∞ optimal design approaches are extended to take into account the cost associated with changes in control inputs. It is shown how to consider the state, transition and input constraints in the design procedure. In the case of exact tracking, the forbidden states, transitions and inputs are avoided by modifying the structural matrices of the system. In the case of optimal tracking, the weight factors in the performance indices are adjusted to penalize the states, transitions and inputs that should be avoided. A numerical example is given to illustrate the results.

In chapter 6, observer-based fault diagnosis (i.e. fault detection and isolation) of BCNs is presented. For passive fault detection, a necessary and sufficient condition for passive fault detectability analysis is introduced. After that, an approach is introduced to design a reduced-order observer for fault detection based on observability analysis. For active fault detection, the concept of active fault detectability is introduced. Then, it is shown that the active fault detection problem of BCNs can be reformulated as a dead-beat stabilization problem of autonomous switched linear discrete-time systems (ASLSs). Based on this, a necessary and sufficient condition is given for active fault detectability analysis. After that, an approach is proposed to design an input sequence generator and an active fault detector. The key is to apply exhaustive search to find many shorter subsequences instead of the longer sequence. If fault occurrence has been detected, then an approach is proposed to solve fault isolation problem of BCNs. The basic idea is to separate non identical dynamics of the faults into different independent subsystems by using graph theory. For each subsystem, a residual generator is constructed based on a reduced-order observer by getting rid of the indistinguishable states.

Chapter 7 gives the summary of the results as well as suggestions for future work.

2 Matrix Expression of Boolean Control Networks

Boolean control networks (BCNs) can be expressed in a matrix form (Cheng et al., 2011). This form enables the study of control problems of BCNs. In order to understand the technique and approaches introduced later, this chapter aims to recall the method introduced in Cheng et al. (2011) to express a Boolean control network (BCN) in a bilinear form similar to standard discrete-time state-space model. In Section 2.1 a mathematical tool called semi-tensor product (STP) of matrices will be presented. Section 2.2 will give an overview of the properties of STP, which are applied to the matrix expression of BCN. Finally, it will be shown in Section 2.3 how to use STP to convert BCNs into a matrix form.

2.1 Semi-Tensor Product of Matrices

In linear algebra, a linear function is a polynomial of degree at most one. As is well known, a linear function can be expressed as the inner product of two vectors. Similarly, a quadratic function is a polynomial function, whose terms have the highest degree of two. In order to express the quadratic function, a quadratic matrix can be used. However, the conventional matrix multiplication can not be used to express polynomials with a degree higher than two. In order to solve the problem, a new matrix multiplication called semi-tensor product (STP) of matrices was firstly introduced by Cheng and co-workers (Cheng et al., 2011), which is a generalized matrix multiplication.

Let A and B be, respectively, a $m \times n$ dimensional matrix and a $p \times q$ dimensional matrix. The Kronecker product of the matrices A and B (also called tensor product) is

Z. Zhang, *Observer Design for Control and Fault Diagnosis of Boolean Networks*,
https://doi.org/10.1007/978-3-658-35929-4_2

$$A \otimes B := \begin{bmatrix} A_{1,1} \cdot B & A_{1,2} \cdot B & \cdots & A_{1,n} \cdot B \\ A_{2,1} \cdot B & A_{2,2} \cdot B & \cdots & A_{2,n} \cdot B \\ \vdots & \vdots & \vdots & \vdots \\ A_{m,1} \cdot B & A_{m,2} \cdot B & \cdots & A_{m,n} \cdot B \end{bmatrix} \tag{2.1}$$

where $\cdot$ denotes matrix multiplication.

Definition 2.1 (Cheng et al. (2011)). The STP of two matrices $A \in \mathbb{R}^{m \times n}$ and $B \in \mathbb{R}^{p \times q}$ is defined as

$$A \ltimes B = (A \otimes I_{l/n}) \cdot (B \otimes I_{l/p})$$

where $\otimes$ denotes the Kronecker product and the $l = lcm\{n, p\}$ is the least common multiple of n and p.

2.2 Properties of the Semi-Tensor Product

The following section aims to outline some fundamental properties of STP introduced in Cheng et al. (2011) and Cheng et al. (2012). The fundamental properties will be used in the subsequent chapters.

The conventional matrix multiplication has some well-known properties. All the properties can still hold by using the generalized matrix multiplication STP. For instance, the transpose of a product of matrices is the product, in the reverse order, of the transposes of the matrices. By applying STP, the following result can be held.

Proposition 2.2 (Cheng et al. (2011)). For matrices A and B, it holds

$$(A \ltimes B)^{\mathrm{T}} = B^{\mathrm{T}} \ltimes A^{\mathrm{T}}. \tag{2.2}$$

However, there are some properties that are not satisfied generally by the conventional matrix multiplication, but by multiplying two or more numbers. For instance, commutativity does not hold by the conventional matrix multiplication, i.e.

$$A \cdot B \neq B \cdot A. \tag{2.3}$$

As a generalized matrix product, the STP brings the significant advantage to have some "commutative" properties with the help of some auxiliary tools. Let δ_n^i denote the i-th column of the identity matrix I_n of the dimensions $n \times n$. The swap matrix is the key tool for pseudo-commutativity defined as follows:

Definition 2.3 (Cheng et al. (2011)). A swap matrix $W_{[m,n]}$ is a $m \cdot n \times m \cdot n$ dimensional matrix given as

$$W_{[m,n]} = \begin{bmatrix} I_m \otimes (\delta_n^1)^{\mathrm{T}} \\ I_m \otimes (\delta_n^2)^{\mathrm{T}} \\ \vdots \\ I_m \otimes (\delta_n^n)^{\mathrm{T}} \end{bmatrix} = \begin{bmatrix} I_n \otimes \delta_m^1 \; I_n \otimes \delta_m^2 \; \cdots \; I_n \otimes \delta_m^m \end{bmatrix}. \tag{2.4}$$

For example, if $m = 3$ and $n = 4$, then the swap matrix $W_{[3,4]}$ is as follows:

$$W_{[3,4]} = \begin{bmatrix} 1\,0\,0\,0\,0\,0\,0\,0\,0\,0\,0\,0 \\ 0\,0\,0\,0\,1\,0\,0\,0\,0\,0\,0\,0 \\ 0\,0\,0\,0\,0\,0\,0\,0\,1\,0\,0\,0 \\ 0\,1\,0\,0\,0\,0\,0\,0\,0\,0\,0\,0 \\ 0\,0\,0\,0\,0\,1\,0\,0\,0\,0\,0\,0 \\ 0\,0\,0\,0\,0\,0\,0\,0\,0\,1\,0\,0 \\ 0\,0\,1\,0\,0\,0\,0\,0\,0\,0\,0\,0 \\ 0\,0\,0\,0\,0\,0\,1\,0\,0\,0\,0\,0 \\ 0\,0\,0\,0\,0\,0\,0\,0\,0\,0\,1\,0 \\ 0\,0\,0\,1\,0\,0\,0\,0\,0\,0\,0\,0 \\ 0\,0\,0\,0\,0\,0\,0\,1\,0\,0\,0\,0 \\ 0\,0\,0\,0\,0\,0\,0\,0\,0\,0\,0\,1 \end{bmatrix}.$$

By using the swap matrix defined in Definition 2.3, the following pseudo-commutativity properties can be obtained.

Proposition 2.4 (Cheng et al. (2011)). Let X and Y be, respectively, a n-dimensional and m-dimensional column vectors. Then, one can get

$$X \ltimes Y = W_{[m,n]} \ltimes Y \ltimes X. \tag{2.5}$$

Proposition 2.5 (Cheng et al. (2011)). Given a $m \times n$ dimensional matrix $A \in \mathbb{R}^{m \times n}$. Let Z be a t-dimensional column vector. Then, there is

$$Z \ltimes A = W_{[m,t]} \ltimes A \ltimes W_{[t,n]} \ltimes Z = (I_t \otimes A) \ltimes Z. \tag{2.6}$$

As a special case, if a STP expression is a product of different Boolean vectors, then one other advantage of STP is that the STP expression can be simplified with the help of a tool called power-reducing matrix Φ_n defined as follows.

Definition 2.6 (Cheng et al. (2011)). A power-reducing matrix Φ_n is a $n^2 \times n$ dimensional matrix given as

$$\Phi_n = \left[\, \delta_n^1 \otimes \delta_n^1 \;\; \delta_n^2 \otimes \delta_n^2 \; \cdots \; \delta_n^n \otimes \delta_n^n \,\right]. \tag{2.7}$$

For example, if $n = 2$, then

$$\Phi_2 = [\delta_2^1 \otimes \delta_2^1 \;\; \delta_2^2 \otimes \delta_2^2] = \delta_4[1 \;\; 4] = \begin{bmatrix} 1 & 0 \\ 0 & 0 \\ 0 & 0 \\ 0 & 1 \end{bmatrix}.$$

By using the power-reducing matrix Φ_n, Cheng et al. (2011) shows that the powers of any Boolean vector in any expression can be reduced to one.

Proposition 2.7 (Cheng et al. (2011)). Let $X \in \mathcal{D}^n$ be a $n \times 1$ dimensional Boolean vector. Then, one can get

$$X \ltimes X = \Phi_n \ltimes X. \tag{2.8}$$

Moreover, STP can also be used to express the other matrix product. For instance, the Hadamard product of two matrices (also called element-wise multiplication) defined in Definition 2.8 can be expressed in (2.10).

Definition 2.8 (Cheng et al. (2012)). Let A and B be two $m \times n$ dimensional matrices. The Hadamard product of A and B is defined as

$$A \odot B = \begin{bmatrix} A_{1,1} \cdot B_{1,1} & A_{1,2} \cdot B_{1,2} & \cdots & A_{1,n} \cdot B_{1,n} \\ A_{2,1} \cdot B_{2,1} & A_{2,2} \cdot B_{2,2} & \cdots & A_{2,n} \cdot B_{2,n} \\ \vdots & \vdots & \ddots & \vdots \\ A_{m,1} \cdot B_{m,1} & A_{m,2} \cdot B_{m,2} & \cdots & A_{m,n} \cdot B_{m,n} \end{bmatrix}. \tag{2.9}$$

Proposition 2.9 (Cheng et al. (2012)). Let A and B be two $m \times n$ dimensional matrices. Then, there is

$$A \odot B = H_m^{\mathrm{T}} \ltimes (A \otimes B) \ltimes H_n \tag{2.10}$$

where the matrix H_n is defined as $H_n = \mathrm{diag}(\delta_n^1, \delta_n^2, \cdots, \delta_n^n)$.

Consider the special case that the element-wise multiplication of vectors should be expressed in a STP form. It is found out that the expression (2.10) can be simplified as follows.

Lemma 2.10 The element-wise multiplication of vectors $X \in \mathbb{R}^{n\times 1}$ and $Y \in \mathbb{R}^{n\times 1}$ can be described by using the transpose of the power-reducing matrix Φ_n^{T} as

$$X \odot Y = \Phi_n^{\mathrm{T}} \ltimes X \ltimes Y. \tag{2.11}$$

Proof. For two column vectors $X, Y \in \mathbb{R}^{n\times 1}$ according to (2.10) in Proposition 2.9 and Definition 2.1, the expression (2.10) can be written as $X \odot Y = H_n^{\mathrm{T}} \ltimes (X \otimes Y) = H_n^{\mathrm{T}} \ltimes X \ltimes Y$. Recall that the matrix H_n is defined as $H_n = diag(\delta_n^1, \delta_n^2, \cdots, \delta_n^n) \in \mathbb{R}^{n^2\times n}$. As the power-reducing matrix Φ_n is explicitly given in (2.7), it can be shown that $H_n = \Phi_n$. Therefore, (2.11) is obtained. □

According to (2.7), it is clear that the power-reducing matrix is actually a logical matrix (i.e each column of the matrix Φ_n contains only one non-zero element 1). In addition, based on the result given in Lemma 2.10, if $X \in \mathcal{D}^n$ is n-dimensional Boolean vector, one can draw the following conclusion.

Lemma 2.11 Consider a Boolean vector $X \in \mathcal{D}^n$. Then one gets

$$X = \Phi_n^{\mathrm{T}} \ltimes X \ltimes X \text{ and } \Phi_n^{\mathrm{T}} \cdot \Phi_n = I_n. \tag{2.12}$$

Proof. Due to the Boolean vector $X \in \mathcal{D}^n$, it can be obtained that $X \odot X = X$. According to (2.11) and letting $Y = X$, then there is $X = X \odot X = \Phi_n^{\mathrm{T}} \ltimes X \ltimes X$. Recall that the i-th column of the power-reducing matrix Φ_n is $\delta_n^i \otimes \delta_n^i = \delta_n^i \ltimes \delta_n^i$. As δ_n^i is a Boolean vector, the expression $\Phi_n^{\mathrm{T}} \cdot \Phi_n = I_n$ holds. □

Lemma 2.11 tells us that with the help of the power-reducing matrix Φ_n, the vector X in any expression can be duplicated.

2.3 Description of Boolean Control Networks

The main purpose of this section is to show the matrix expression of Boolean control networks (BCNs), which was proposed in Cheng and Qi (2010) by using STP.

A BCN can be represented by the following equations (Cheng et al., 2011; Cheng et al., 2015):

$$X_i(t+1) = f_i(X(t), U(t)), \qquad i = 1, 2, \cdots, n, \tag{2.13}$$

$$Y_j(t) = h_j(X(t)), \qquad j = 1, 2, \cdots, p, \tag{2.14}$$

where $X(t) = [X_1(t)\, X_2(t) \cdots X_n(t)]^{\mathrm{T}} \in \mathcal{D}^n, U(t) = [U_1(t)\, U_2(t) \cdots U_m(t)]^{\mathrm{T}} \in \mathcal{D}^m$ and $Y(t) = [Y_1(t)\ Y_2(t) \ \cdots \ Y_p(t)]^{\mathrm{T}} \in \mathcal{D}^p$ are, respectively, the state vector, input vector and output vector at time t, f_i and h_j are Boolean functions $\mathcal{D}^{n \times m} \xrightarrow{f_i} \mathcal{D}$ and $\mathcal{D}^n \xrightarrow{h_j} \mathcal{D}$, respectively.

According to Franke (1995), a Boolean function is a multi-linear function which can be expressed as a polynomial with higher degree (i.e. polynomial form)

$$\begin{aligned} f_{\mathbb{B}}(X) =& a_0 + \sum_{i=1}^{n} a_i \cdot X_i + \sum_{j=2}^{n} \sum_{i=1}^{j-1} a_{ij} \cdot X_i \cdot X_j \\ &+ \sum_{k=3}^{n} \sum_{j=2}^{k-1} \sum_{i=1}^{j-1} a_{ijk} \cdot X_i \cdot X_j \cdot X_k + \cdots + a_{123 \cdots n} \cdot X_1 \cdot X_2 \cdot X_3 \cdot \ \cdots \ \cdot X_n. \end{aligned} \tag{2.15}$$

However, due to the nonlinear terms, e.g. $a_{123\cdots n} \cdot X_1 \cdot X_2 \cdot X_3 \cdot \cdots \cdot X_n$ it is very hard to study BCNs (2.13)–(2.14) in a control theoretical framework. In order to solve the problem, a matrix representation of BCNs has been proposed in Cheng and Qi (2010) by applying STP. This approach converts the dynamics of the BCNs into a bilinear model that is quite similar to the standard discrete-time state-space model. Hence, it enables to study control problems of BCNs.

Before introducing the matrix representation of BCNs, it will be shown that any multi-linear function can be expressed as a sequence of products among vectors by using STP. For the sake of simplicity, consider a bilinear (quadratic) function $f(X, U)$. The domain of the function $f(X, U)$ is defined over the n-dimensional vector space $X = [X_1\ X_2\ \cdots\ X_n]^{\mathrm{T}} \in \mathbb{R}^n$ and m-dimensional vector space $U = [U_1\ U_2\ \cdots\ U_m]^{\mathrm{T}} \in \mathbb{R}^m$. Recall that Δ_n and Δ_m are, respectively, the sets of basis of the spaces X and U. Let a $n \times m$ dimensional of f be denoted by

$$S = \begin{bmatrix} S_{1,1} & S_{1,2} & \cdots & S_{1,m} \\ S_{2,1} & S_{2,2} & \cdots & S_{2,m} \\ \vdots & \vdots & \vdots & \vdots \\ S_{n,1} & S_{n,2} & \cdots & S_{n,m} \end{bmatrix}.$$

with the (i, j)-th element $S_{i,j},\ i = 1, 2, \cdots, n;\ j = 1, 2, \cdots, m$ calculated by

$$S_{i,j} = f(\delta_n^i, \delta_m^j). \tag{2.16}$$

Then the bilinear function $f(X, U)$ can be expressed as

$$f(X, U) = X^{\mathrm{T}} \cdot S \cdot U = \sum_{i=1}^{n} \sum_{j=1}^{m} S_{i,j} \cdot X_i \cdot U_j \tag{2.17}$$

where the matrix S is called structure matrix of the function f. Let the elements $S_{i,j},\ i = 1, 2, \cdots, n;\ j = 1, 2, \cdots, m$ in the matrix S be arranged in a $1 \times n \cdot m$ dimensional vector as follows:

$$V_S = [S_{1,1}\ S_{1,2}\ \cdots\ S_{1,m}\ \ S_{2,1}\ \cdots\ S_{n,1}\ \cdots\ S_{n,m}]. \tag{2.18}$$

When the vector V_s is partitioned into n blocks with the i-th block $Blk_i(V_S) = [S_{i,1}\ S_{i,2}\ \cdots\ S_{i,m}]$, according to Definition 2.1, $V_S \ltimes X \ltimes U$ can be expressed as

$$\begin{aligned} V_S \ltimes X \ltimes U &= \left(\sum_{i=1}^{n} Blk_i(V_S) \cdot X_i \right) \ltimes U \\ &= \left(\sum_{i=1}^{n} [S_{i,1} \cdot X_i\ \ S_{i,2} \cdot X_i\ \ \cdots\ \ S_{i,m} \cdot X_i] \right) \ltimes U \\ &= \sum_{i=1}^{n} \sum_{j=1}^{m} S_{i,j} \cdot X_i \cdot U_j \end{aligned} \tag{2.19}$$

which is actually equal to (2.17). This means that the quadratic function $f(X, U)$ can be expressed by using STP as $V_S \ltimes X \ltimes U$. In a similar manner, the result can easily be extended to a multi-linear function (Cheng et al., 2012). Consider a k-linear function $f : \mathbb{R}^{n_1} \times \mathbb{R}^{n_2} \times \cdots \times \mathbb{R}^{n_k} \to \mathbb{R}$. Let X_i be denoted as $X_i = [X_1^i\ X_2^i\ \cdots\ X_{n_i}^i]^{\mathrm{T}},\ i = 1, 2, \cdots, k$. The k-linear function $f(X_1, X_2, \cdots, X_k)$ takes the following form:

$$f(X_1, X_2, \cdots, X_k) = \sum_{i_1=1}^{n_1} \sum_{i_2=1}^{n_2} \cdots \sum_{i_k=1}^{n_k} S_{i_1, i_2, \cdots, i_k} \cdot X_{i_1}^1 \cdot X_{i_2}^2 \cdot \quad \cdots \quad \cdot X_{i_k}^k. \quad (2.20)$$

Let the coefficients $S_{i_1, i_2, \cdots, i_k}$ be arranged in a vector V_S according to the order that $i_t, t = k$ runs from 1 to n_t and then let the index i_{t-1} go from 1 to n_{t-1} and so on. At the end, the k-linear function $f(X_1, X_2, \cdots, X_k)$ can be expressed as

$$f(X_1, X_2, \cdots, X_k) = V_S \ltimes X_1 \ltimes X_2 \ltimes \cdots \ltimes X_k. \quad (2.21)$$

As a multi-linear function, a Boolean function can be formulated into the matrix expression (2.21). For this purpose, let the vectors δ_2^1 and δ_2^2 represent, respectively, the logic "true" and "false". Accordingly, as a Boolean variable X can take value in the set $\mathcal{D}$, let the variable X be expressed in a vector form x as described in Fornasini and Valcher (2013):

$$x = \begin{bmatrix} X & \neg X \end{bmatrix}^{\mathrm{T}} \quad (2.22)$$

where the superscript T denotes the transpose.

Next, system matrix of a BCN (2.13)–(2.14) will be constructed which contains all the structural information on the logical functions $f_i, i = 1, 2, \cdots, n$ and $h_j, j = 1, 2, \cdots, p$. For this purpose, Cheng et al. (2011) pointed out that for any logical operator, a structure matrix can be found to express any logical operator. For instance, the structure matrix, denoted by M_n, for the logical operator $\neg$ is

$$M_n = \begin{bmatrix} 0 & 1 \\ 1 & 0 \end{bmatrix} = \delta_2[2 \ \ 1]. \quad (2.23)$$

If a Boolean variable X is expressed in a vector form x according to (2.22), then one obtains

$$M_n \ltimes x = \begin{bmatrix} 0 & 1 \\ 1 & 0 \end{bmatrix} \cdot \begin{bmatrix} X \\ \neg X \end{bmatrix} = \begin{bmatrix} \neg X \\ X \end{bmatrix} \quad (2.24)$$

which is the vector form of the Boolean variable $\neg X$. Similarly, for conjunction, disjunction, conditional and biconditional, the corresponding strucutre matrices are

$$\begin{aligned} \text{Conjunction}: \quad & M_c = \delta_2[1\ 2\ 2\ 2], & (2.25)\\ \text{Disjunction}: \quad & M_d = \delta_2[1\ 1\ 1\ 2], & (2.26)\\ \text{Conditional}: \quad & M_i = \delta_2[1\ 2\ 1\ 1], & (2.27)\\ \text{Biconditional}: \quad & M_e = \delta_2[1\ 2\ 2\ 1]. & (2.28) \end{aligned}$$

Denote x_i as the vector form of the Boolean variables X_i. Then the following result can be obtained.

Lemma 2.12 (Cheng et al. (2011)). Given a Boolean function $f(X_1, X_2, \cdots, X_n)$. There exists a unique 2×2^n dimensional structure matrix M_f of the Boolean function f, such that the Boolean function $f(X_1, X_2, \cdots, X_n)$ can be represented by

$$f(X_1, X_2, \cdots, X_n): \ M_f \ltimes x_1 \ltimes x_2 \ltimes \cdots \ltimes x_n. \tag{2.29}$$

Example 2.13 Assume that a Boolean function $f(X_1, X_2, X_3) = (X_1 \vee X_2) \wedge (X_2 \vee X_3)$ is given. Recall that the logical operator "conjunctive" (or "disjunction") can be expressed in matrix form by using the structure matrix M_c (or M_d) defined in (2.26) (or in (2.25)). Then, there is

$$f(X_1, X_2, X_3) := M_d \ltimes (M_c \ltimes x_1 \ltimes x_2) \ltimes (M_c \ltimes x_2 \ltimes x_3). \tag{2.30}$$

Applying the properties of STP given in Proposition 2.4–2.7, (2.30) can be further simplified as

$$\begin{aligned} f(X_1, X_2, X_3) : &= M_d \ltimes M_c \ltimes (I_4 \otimes M_c) \ltimes x_1 \ltimes x_2 \ltimes x_2 \ltimes x_3 \\ &= M_d \ltimes M_c \ltimes (I_4 \otimes M_c) \ltimes x_1 \ltimes \Phi_2 \ltimes x_2 \ltimes x_3 \\ &= \underbrace{M_d \ltimes M_c \ltimes (I_4 \otimes M_c) \ltimes (I_2 \otimes \Phi_2)}_{M_f} \ltimes x_1 \ltimes x_2 \ltimes x_3. \end{aligned}$$

Let $x(t) = x_1(t) \ltimes x_2(t) \ltimes \cdots \ltimes x_n(t) \in \Delta_{2^n}$, $u(t) = u_1(t) \ltimes u_2(t) \ltimes \cdots \ltimes u_m(t) \in \Delta_{2^m}$ and $y(t) = y_1(t) \ltimes y_2(t) \ltimes \cdots \ltimes y_p(t) \in \Delta_{2^p}$. According to Lemma 2.12, the Boolean functions in (2.13) can be expressed as

$$x_i(t+1) = M_{f_i} \ltimes u(t) \ltimes x(t). \tag{2.31}$$

Based on this, the state $x(t+1) = x_1(t+1) \ltimes x_2(t+1) \ltimes \cdots \ltimes x_n(t+1)$ can be represented as

$$\begin{aligned}
x(t+1) &= x_1(t+1) \ltimes x_2(t+1) \ltimes \cdots \ltimes x_n(t+1) \\
&= M_{f_1} \ltimes u(t) \ltimes x(t) \ltimes M_{f_2} \ltimes u(t) \ltimes x(t) \ltimes \cdots \ltimes M_{f_n} \ltimes u(t) \ltimes x(t) \\
&= M_{f_1} \ltimes (I_{2^{n+m}} \otimes M_{f_2}) \ltimes \Phi_{2^{n+m}} \ltimes u(t) \ltimes x(t) \ltimes \cdots \ltimes M_{f_n} \ltimes u(t) \ltimes x(t) \\
&= \underbrace{\left(\prod_{i=1}^{n} \left((I_{2^{(i-1)\cdot(n+m)}} \otimes M_{f_i}) \ltimes \Phi_{2^{n+m}}\right)\right)}_{L} \ltimes u(t) \ltimes x(t).
\end{aligned}$$

Similarly, the output $y(t)$ can be calculated by $y(t) = H \ltimes x(t)$. As a result, the BCN (2.13)–(2.14) can be equivalently written as:

$$x(t+1) = L \ltimes u(t) \ltimes x(t), \tag{2.32}$$

$$y(t) = H \ltimes x(t), \tag{2.33}$$

where $L \in \mathcal{L}_{2^n \times 2^{n+m}}$ and $H \in \mathcal{L}_{2^p \times 2^n}$ are the logical matrices that contain all the structural information on the logical functions. One can also use swap matrix $W_{[2^n, 2^m]}$ to reformulate (2.32) equivalently. Let the matrix $L_{eq} = LW_{[2^n, 2^m]}$. The BCN (2.13)–(2.14) can also be rewritten as

$$x(t+1) = L_{eq} \ltimes x(t) \ltimes u(t), \tag{2.34}$$

$$y(t) = H \ltimes x(t). \tag{2.35}$$

It is worth pointing out that by considering the BCN in the form of (2.34)–(2.35), the state observers introduced later will be expressed in a more compact form then using the form (2.32)–(2.33).

2.4 Boolean Matrices

Let $\mathcal{D} \in \{0, 1\}$. The definition of Boolean matrices given in Cheng et al. (2011) will be recalled.

Definition 2.14 A matrix M of dimensions $m \times n$ is called a Boolean matrix, if all the entries in the matrix M belong to the set $\mathcal{D}$, i.e. $M_{i,j} \in \mathcal{D}$, $\forall i = 1, 2, \cdots, m$; $j = 1, 2, \cdots, n$.

In some applications (for instance, reconstructibility analysis, observer design, etc.), it is merely required to know whether the entry of a matrix is positive or not. For this, the entry of the matrix can be expressed by using the truth values, *true* and *false*, i.e. elements in the set $\mathcal{D}$. That means the matrix (vector) can be simply converted into a Boolean matrix (vector). Besides, Boolean algebra can be used in the calculation. Assume that $Z = A \ltimes B$. The Boolean product is defined in Cheng et al. (2011) as

$$A \ltimes_{\mathcal{B}} B := C, \text{ with } C_{i,j} = \begin{cases} 0, & Z_{i,j} = 0 \\ 1, & Z_{i,j} \neq 0 \end{cases}, \forall i, j. \tag{2.36}$$

For instance, let the matrices A and B be, respectively,

$$A = \begin{bmatrix} 1\ 1 \\ 0\ 0 \\ 0\ 0 \\ 0\ 1 \end{bmatrix} \text{ and } B = \begin{bmatrix} 5 \\ 6 \end{bmatrix}.$$

The Boolean product of the matrices A and B is

$$A \ltimes_{\mathcal{B}} B = \begin{bmatrix} 1\ 1 \\ 0\ 0 \\ 0\ 0 \\ 0\ 1 \end{bmatrix} \ltimes_{\mathcal{B}} \begin{bmatrix} 5 \\ 6 \end{bmatrix} = \begin{bmatrix} 1 \\ 0 \\ 0 \\ 1 \end{bmatrix}. \tag{2.37}$$

In contrast to that, the STP of the matrices A and B is

$$A \ltimes B = \begin{bmatrix} 1\ 1 \\ 0\ 0 \\ 0\ 0 \\ 0\ 1 \end{bmatrix} \ltimes \begin{bmatrix} 5 \\ 6 \end{bmatrix} = \begin{bmatrix} 11 \\ 0 \\ 0 \\ 6 \end{bmatrix}. \tag{2.38}$$

After comparing (2.37) to (2.38), it can be seen that $A \ltimes_{\mathcal{B}} B$ indicates the non-zero entries in $A \ltimes B$. It is important to point out that if A and B are logical matrices or vectors, then $A \ltimes_{\mathcal{B}} B = A \ltimes B$.

In the following parts, depending on different applications, the Boolean product or the STP will be used. In order to reduce notational overdose, the notation $\ltimes_{\mathcal{B}}$ for the Boolean productor or $\ltimes$ for STP will be omitted overall in the thesis. At beginning of each chapter, it will be pointed out explicitly, which operator is used.

3 Reconstructibility analysis

The objective of this chapter is to introduce an approach for reconstructibility analysis of BCNs. The relationship between the various definitions of observability and reconstructibility of BCNs is given at first. For reconstructibility analysis of BCNs, an explicit method by applying STP is introduced. To facilitate the application to larger networks, a recursive method for reconstructibility analysis is derived.

For a simple derivation of results, in this chapter BCNs in the form (2.34)–(2.35) will be considered and the Boolean product (2.36) will be applied. For the sake of simplicity, the symbol $\ltimes_{\mathcal{B}}$ will be omitted.

3.1 Observability and Reconstructibility

In this section, the aim is to give a clear picture about the relationship between the various definitions of observability and reconstructibility. The revealed result is very important for the observer design introduced later.

3.1.1 Observability

In reality, not all state variables of BCNs can be measured directly. The property "observability" is introduced to measure whether the initial state of the BCNs can be inferred with the knowledge of the input and output trajectory. In the literature, different definitions of observability for BCNs can be found.

Definition 3.1 (Cheng and Qi, 2009) A BCN is said to be observable if for any initial state $x(0)$ there exists an input sequence $\{u(0), u(1), \cdots\}$, such that for any $\bar{x}(0) \neq x(0)$ the corresponding output sequences $\{\bar{y}(0), \bar{y}(1), \cdots\} \neq \{y(0), y(1), \cdots\}$.

Z. Zhang, *Observer Design for Control and Fault Diagnosis of Boolean Networks*, https://doi.org/10.1007/978-3-658-35929-4_3

Definition 3.2 (Zhao et al., 2010) A BCN is said to be observable if for any two distinct states $\bar{x}(0)$, $x(0)$ there is an input sequence $\{u(0), u(1), \cdots, u(r)\}, r \in \mathbb{Z}_+$, such that the corresponding output sequences $\{\bar{y}(0), \bar{y}(1), \cdots, \bar{y}(r)\} \neq \{y(0), y(1), \cdots, y(r)\}$.

Definition 3.3 (Laschov et al., 2013) A BCN is said to be observable if there exists an input sequence $\{u(0), u(1), \cdots, u(r)\}, r \in \mathbb{Z}_+$, such that for any two distinct $\bar{x}(0)$, $x(0)$, the corresponding output sequences $\{\bar{y}(0), \bar{y}(1), \cdots, \bar{y}(r)\} \neq \{y(0), y(1), \cdots, y(r)\}$.

Definition 3.4 (Fornasini and Valcher, 2013) A BCN is said to be observable if for any two distinct states $\bar{x}(0)$, $x(0)$ and any input sequence $\{u(0), u(1), \cdots\}$, the corresponding output sequences $\{\bar{y}(0), \bar{y}(1), \cdots\} \neq \{y(0), y(1), \cdots\}$.

As BCNs are indeed nonlinear systems, Cheng and Qi (2009) as well as Zhao et al. (2010) propose the observability (i.e. Definition 3.1 and Definition 3.2) that relay on the initial states and input sequences. In order to study identification problems of BCNs, Definition 3.3 is proposed as a condition for identifiability. Like the observability of linear systems (i.e. for all sufficiently long input sequences), observability in Definition 3.4 is given in Fornasini and Valcher (2013).

Recently, the relationship between different definitions of observability has been studied in Zhang and Zhang (2016). The upper part of Fig. 3.1 shows the relationship among Definition 3.1-Definition 3.4 (Zhang and Zhang, 2016). From Fig. 3.1 one gets that if a BCN is observable in the sense of Definition 3.4, then it is also observable in the sense of Definition 3.1-Definition 3.3.

3.1.2 Reconstructibility

In the case of reconstructibility, it is assumed that the current state should be determined based on the knowledge of the past and current input and output measurements. Similar to observability, various types of reconstructibility have been proposed for different situations. It is important to note that current state observability introduced in Xu and Hong (2013) is defined in a similar manner as reconstructibility. Therefore, current state observability defined in Xu and Hong (2013) will be considered as a kind of reconstructibility.

Definition 3.5 (Xu and Hong, 2013) A BCN is said to be reconstructible, if there is a positive integer r, such that for arbitrary unknown initial state $x(0)$, the current state $x(r+1)$ can be uniquely determined from every admissible[1] input and output trajectory $\{(y(t), u(t)), t = 0, 1, \cdots, r\}, r \in \mathbb{N}$.

Definition 3.6 (Zhang et al., 2016) A BCN is said to be reconstructible, if there is an input sequence $\{u(0), u(1), \cdots, u(r)\}, r \in \mathbb{N}$, such that for all different initial states $x(0)$ and $\bar{x}(0)$, $x(r) \neq \bar{x}(r)$ implies $\{\bar{y}(0), \bar{y}(1), \cdots, \bar{y}(r)\} \neq \{y(0), y(1), \cdots, y(r)\}$.

Definition 3.7 (Fornasini and Valcher, 2013) with the knowledge of every admissible input and output trajectory $\{(y(t), u(t)), t = 0, 1, \cdots, r\}, r \in \mathbb{N}$ the final state $x(r)$ can be uniquely determined.

Note that the reconstructibility in the sense of Definition 3.5 is equivalent to Definition 3.7. Therefore, Definition 3.5 will not be considered separately.
For fault diagnosis, Sridharan et al. (2012) applied an input sequence to drive BCNs to a known state independent of initial states. Based on this, reconstructibility in Definition 3.6 is introduced in Zhang et al. (2016). Definition 3.7 is proposed to be consistent with the requirement of state estimation for every sufficiently long input sequence.

The relationship between Definition 3.6 and Definition 3.7 will be studied. Definition 3.6 is not the most restrictive as it does not require that the final state can be inferred from the knowledge of every admissible input and output trajectory. Therefore, if a BCN is reconstructible in the sense of Definition 3.7, then the BCN is also reconstructible in the sense of Definition 3.6. But the opposite is not always true.

3.1.3 Relationships Between Observability and Reconstructibility

In this part, the relationship between observability and reconstructibility shall be studied. It will be shown that the observability is not equivalent to the reconstructibility in general. The observability in the sense of some definitions implies directly the reconstructibility. But it does not hold vice versa.

[1] It is worthwhile to mention that an input and output trajectory is admissible, if there exists an initial state $x(0) \in \Delta_{2^n}$ such that applying the input sequence $\{u(0), u(1), \cdots, u(t)\}$ the corresponding output sequence is $\{y(0), y(1), \cdots, y(t)\}$ (Fornasini and Valcher, 2013).

At first, the observability in the sense of Definition 3.4 and the reconstructibility in the sense of Definition 3.7 are considered. After comparing the definitions, the following result can be obtained.

Theorem 3.8 *If a BCN described by (2.34)–(2.35) is observable in the sense of Definition 3.4, then the BCN is reconstructible in the sense of Definition 3.7.*

Proof. If an input $u(t)$ is given, then the successor state $x(t+1)$ of the state $x(t)$ is uniquely determined according to (2.34). After recursive applying (2.34), there is

$$x(r) = L_{eq}^{r} x(0)u(0)u(1)\cdots u(r-1). \tag{3.1}$$

Once the knowledge of an input and output trajectory $\{(y(t), u(t)), t = 0, 1, \cdots, r\}$ allows to determine the initial state $x(0)$, the state $x(r)$ can also be uniquely determined according to (3.1) and the input sequence $\{u(0)u(1)\cdots u(r-1)\}$. According to Definition 3.7, the BCN is reconstructible. □

For observability analysis in the sense of Definition 3.4, Fornasini and Valcher have given two conditions in Fornasini and Valcher (2013) that should be checked, i.e. distinguishability of states before state merging and distinguishability of states belonging to cycles (see Theorem 3 in Fornasini and Valcher (2013) for detail). Different from that, in order to check reconstructibility in the sense of Definition 3.7 only the condition "distinguishability of states belonging to cycles" should be satisfied (see Theorem 4 in Fornasini and Valcher (2013)). Therefore, generally the observability of BCNs in the sense of Definition 3.4 is stricter than the reconstructibility of BCNs in the sense of Definition 3.7. Only in some special cases, the observability in the sense of Definition 3.4 and the reconstructibility in the sense of Definition 3.7 are equivalent. One of these cases is described in Theorem 3.9.

Theorem 3.9 *Consider a BCN described by (2.34)–(2.35). The observability analysis based on Definition 3.4 and the reconstructibility analysis based on Definition 3.5 are equivalent, if the matrix Ξ calculated by $\Xi = L_{eq}H^{\mathrm{T}}$ holds*

$$\sum_{i=1}^{2^n} \Xi_{i,j} = \sum_{i=1}^{2^n} H_{i,j}^{\mathrm{T}}, \ \forall j = 1, 2, \cdots, 2^{p+m}. \tag{3.2}$$

Proof. Suppose that each column of the matrix M of dimensions $2^p \times 2^n$ belongs to the set $\Delta_{2^p} \cup \{\mathbf{0}_{2^p}\}$ in any equations like $y = Mx$. Then $x = \delta_{2^n}^{i}$ is a solution for the equation $\delta_{2^p}^{j} = Mx$, if and only if $M_{j,i} \neq 0$. This is equivalent to $\left[M^{\mathrm{T}}\right]_{i,j} \neq 0$.

Hence the vector $\hat{x}$ containing all the possible solutions for the equation $y = Mx$ is obtained by

$$\hat{x} = M^{\mathrm{T}} y \tag{3.3}$$

and $\delta_{2^n}^i$ is a solution as long as $[\hat{x}]_i \neq 0$.

Similarly, it can be derived from (2.35) that for an output $y \in \Delta_{2^p}$ the vector $H^{\mathrm{T}} y$ represents a set of states that can generate the output y. Recall that any column vector in the set Δ_{2^p} contains only one non-zero entry 1. Therefore, each column of the matrix H^{T} corresponds to the result of one possible output, i.e. $[H^{\mathrm{T}}\delta_{2^p}^1 \;\; H^{\mathrm{T}}\delta_{2^p}^2 \;\; \cdots \;\; H^{\mathrm{T}}\delta_{2^p}^{2^p}] = H^{\mathrm{T}}$. According to (2.34), the states reached in one step can be calculated by $L_{eq} H^{\mathrm{T}}$. If there exists an index $j \in \{1, 2, \cdots, 2^{p+m}\}$, such that $\sum_{i=1}^{2^n} \Xi_{i,j} < \sum_{i=1}^{2^n} H_{i,j}^{\mathrm{T}}$, then this implies that two states $x_1 \neq x_2$ with $x_1, x_2 \in \Delta_{2^n}$ and an input $u \in \Delta_{2^m}$ exist, such that $L_{eq} x_1 u = L_{eq} x_2 u$ and $H x_1 = H x_2$. Then, according to i) of Theorem 3 in Fornasini and Valcher (2013), the BCN is not observable. □

Note that if condition (3.2) is satisfied, then the reconstructibility in the sense of Definition 3.7 is also sufficient for the observability in the sense of Definition 3.1-Definition 3.3. But due to the dependence of the input sequence, it can be clearly concluded that the observability in the sense of Definition 3.1-Definition 3.3 does not imply reconstructibility in the sense of Definition 3.7 in general.

In the following, the relationship between the observability in the sense of Definition 3.3 and the reconstructibility in the sense of Definition 3.6 will be clarified. By comparing the definitions, the following conclusion can be obtained which is similar to Theorem 3.8.

Theorem 3.10 *If a BCN described by (2.34)–(2.35) is observable in the sense of Definition 3.3, then the BCN is reconstructible in the sense of Definition 3.6.*

As a matter of fact, the observability in the sense of Definition 3.3 can imply the reconstructibility in the sense of Definition 3.6. But the opposite is not always true. As a simple counterexample, the following BCN is considered

$$\begin{aligned} x(t+1) &= \delta_4[3\ 2\ 3\ 2\ 2\ 3\ 4\ 1]u(t)x(t), \\ y(t) &= \delta_2[1\ 2\ 1\ 2]x(t), \end{aligned} \tag{3.4}$$

where $x \in \Delta_4$, $u \in \Delta_2$, $y \in \Delta_2$. It is simply to verify that this BCN is controllable. Beginning with the output $y(0) \in \Delta_2$, the possible initial states belong to the set $\{\delta_4^1, \delta_4^3\}$ or $\{\delta_4^2, \delta_4^4\}$. If the input sequence $\{u(t) = \delta_2^2, t = 0, 1, \cdots\}$ is used, then the transition of the set of states is $\{\delta_4^1, \delta_4^3\} \xrightarrow{u=\delta_2^2} \{\delta_4^2, \delta_4^4\} \xrightarrow{u=\delta_2^2} \{\delta_4^1, \delta_4^3\} \xrightarrow{u=\delta_2^2} \cdots$. Moreover, it is easy to understand that when the input $u = \delta_2^1$ is applied, the successor states of all the states belonging to the set $\{\delta_4^1, \delta_4^3\}$ and $\{\delta_4^2, \delta_4^4\}$ are, respectively, δ_4^3 and δ_4^2. Therefore, once $u = \delta_2^1$ is adopted, it is impossible to identify the initial state, but the final state can be uniquely determined. □

For the observability in the sense of Definition 3.1 and Definition 3.2, each initial state may be identified by applying different input sequences. In contrast, for the reconstructibility in the sense of Definition 3.6 only one input sequence needs to be applied to distinguish the final states, while the reconstructibility in the sense of Definition 3.7 requires the consideration of every possible input sequence. Hence, there is no relationship between the reconstructibility in Definition 3.5-Definition 3.7 and the observability in the sense of Definition 3.1 and Definition 3.2.

The relations revealed in Section 3.1.2 and 3.1.3 are added to Fig. 3.1 (see the bold solid line and the bold dashed line).

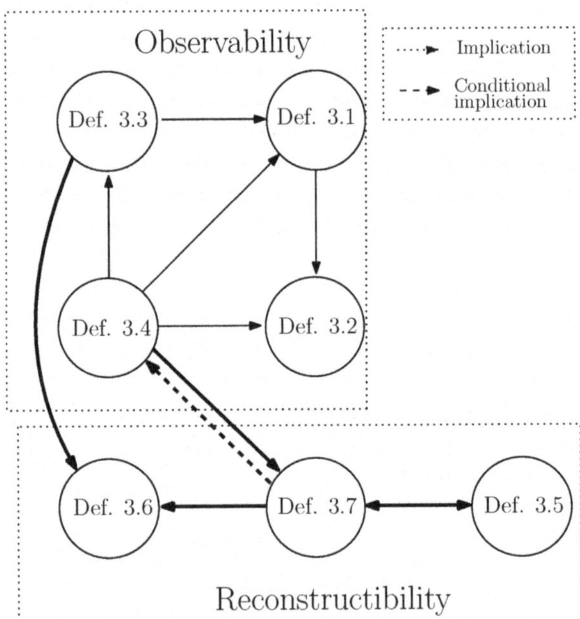

Figure 3.1 The relationships between observability in Definition 3.1-Definition 3.4 and reconstructibility as described in Definition 3.5-Definition 3.7. The bold solid and bold dashed lines show the relations revealed in Section 3.1.2 and 3.1.3

3.2 Reconstructibility Analysis

As pointed out in Fornasini and Valcher (2013), a necessary condition for the existence of a state observer is that the BCN is reconstructible. As shown in Fig. 3.1, the reconstructibility in Definition 3.7 is stronger than the reconstructibility in Definition 3.6. In practice, one wants to get a good state estimation without any limitation on input sequence. However, a good state estimation under the reconstructibility in Definition 3.6 can be provided only if the suitable input sequence is applied. Hence, reconstructibility analysis in the sense of Definition 3.7 will be considered. For this, an explicit method will be proposed by which a function of the input and output trajectory to express the current state will be derived. After that, to avoid high dimensional matrices and to reduce the computational effort, an equivalent recursive method will be proposed. Usually the index r in Definition 3.7 is unknown. Different from the existing approaches, the recursive method also determines the index r and does not require the knowledge of formal language. Besides, a stopping criteria will be given to terminate the recursive method earlier if a BCN is not reconstructible. It is worth pointing out that the problem of reconstructibility analysis of BCNs is generally a $\mathcal{NP}$-hard problem (Zhang et al., 2016).

3.2.1 Explicit Method

According to Definition 3.7 and Lemma 2.11, the following theorem can be obtained.

Theorem 3.11 *A BCN described by (2.34)–(2.35) is reconstructible, if and only if there exists a non-negative integer r, so that there is at most one non-zero entry in each column of the matrix Ω_r of dimensions $2^n \times 2^{r\cdot(p+m)+p}$ calculated by*

$$\Omega_r = (\Phi_{2^n}^{\mathrm{T}} L_{eq})^r \left(\prod_{i=0}^{r} \left(I_{2^{i(n+m)}} \otimes H \right) \right)^{\mathrm{T}}. \tag{3.5}$$

Proof. According to (2.35) and applying the property of STP given in Proposition 2.5, any input and output trajectory of $r+1$ (i.e. $(u(t), y(t))$, $t = 0, 1, \cdots, r$) can be expressed as a function of an input and state trajectory, i.e.

$$\prod_{j=0}^{r} y(j)u(j) = \prod_{i=0}^{r} \left(Hx(i)u(i) \right) = \left(\prod_{i=0}^{r} \left(I_{2^{i\cdot(n+m)}} \otimes H \right) \right) \left(\prod_{i=0}^{r} x(i)u(i) \right). \tag{3.6}$$

Hence, all the possible input and state trajectories that lead to the given input and output trajectory can be obtained by

$$\omega = \left(\prod_{i=0}^{r}\left(I_{2^{i(n+m)}} \otimes H\right)\right)^{\mathrm{T}}\left(\prod_{j=0}^{r} y(j)u(j)\right), \tag{3.7}$$

where ω is a vector of the dimension $2^{(r+1)(m+n)} \times 1$. $x(0)u(0)x(1)u(1)\cdots x(r)u(r) = \delta^i_{2^{(r+1)(n+m)}}$ is one possible input and state trajectory, if and only if $[\omega]_i \neq 0$. Considering (2.34) and applying Lemma 2.11, there is

$$x(1) = L_{eq}x(0)u(0) = L_{eq}\Phi^{\mathrm{T}}_{2^n}x(0)x(0)u(0). \tag{3.8}$$

In the same way, $x(2)$ can be determined by

$$\begin{aligned} x(2) &= L_{eq}\Phi^{\mathrm{T}}_{2^n}x(1)x(1)u(1) \\ &= (L_{eq}\Phi^{\mathrm{T}}_{2^n})^2 x(0)x(0)u(0)x(1)u(1) \\ &= L_{eq}\Phi^{\mathrm{T}}_{2^n}L_{eq}x(0)u(0)x(1)u(1). \end{aligned} \tag{3.9}$$

Applying the above procedure for $t = 0, 1, \cdots, r+1$, the state $x(r+1)$ can be expressed by

$$x(r+1) = \left(L_{eq}\Phi^{\mathrm{T}}_{2^n}\right)^r L_{eq}\left(\prod_{j=0}^{r} x(j)u(j)\right). \tag{3.10}$$

Applying (3.7) and based on (3.10), all possible states $x(r+1)$ along the given input and output trajectory $\{(y(t), u(t)), t = 0, 1, \cdots, r\}$ can be determined as

$$\begin{aligned} \hat{x}(r+1) &= \left(L_{eq}\Phi^{\mathrm{T}}_{2^n}\right)^r L_{eq}\,\omega \\ &= \left(L_{eq}\Phi^{\mathrm{T}}_{2^n}\right)^r L_{eq}\left(\prod_{i=0}^{r}\left(I_{2^{i(n+m)}} \otimes H\right)\right)^{\mathrm{T}}\left(\prod_{j=0}^{r} y(j)u(j)\right) \\ &= L_{eq}\left(\Phi^{\mathrm{T}}_{2^n}\left(L_{eq}\Phi^{\mathrm{T}}_{2^n}\right)^{r-1} L_{eq}\left(\prod_{i=0}^{r}\left(I_{2^{i(n+m)}} \otimes H\right)\right)^{\mathrm{T}}\left(\prod_{j=0}^{r-1} y(j)u(j)\right)y(r)\right)u(r), \end{aligned} \tag{3.11}$$

where $\delta^i_{2^n}$ is a candidate state at time $r+1$, if and only if $\left[\hat{x}(r+1)\right]_i > 0$.

From (3.11), a similar form as (2.34) can be found. Then the states $\hat{x}(r)$ that are compatible with the input and output trajectory $\{y(0), u(0), y(1), u(1), \cdots, y(r)\}$

can be expressed as

$$\hat{x}(r) = \left(\Phi_{2^n}^{\mathrm{T}} L_{eq}\right)^r \left(\prod_{i=0}^{r} \left(I_{2^{i(n+m)}} \otimes H\right)\right)^{\mathrm{T}} \left(\prod_{j=0}^{r-1} y(j)u(j)\right) y(r). \tag{3.12}$$

The state estimate $\hat{x}(r)$ is based on the input and output trajectory $\{y(0), u(0), y(1), u(1), \cdots, y(r-1), u(r-1), y(r)\}$. $\delta_{2^n}^i$ is a candidate state if and only if $\left[\hat{x}(t+r)\right]_i > 0$. From (3.12), it can be obtained that the matrix Ω_r defined by (3.5) has the dimensions $2^n \times 2^{r\cdot(m+p)+p}$ and the columns of Ω_r show the results calculated by different input and output trajectories. If there is no non-zero entry in a column of Ω_r, then the corresponding input and output trajectory is not admissible. If there is only one non-zero entry in Ω_r, then the corresponding input and output trajectory is admissible and the final state can be uniquely determined. Otherwise, the final state can not be uniquely determined. □

The minimal integer r satisfying the condition given in Theorem 3.11 is called the minimal reconstructibility index and will from now on be denoted by r_{min}. The index r_{min} also indicates that the state estimate provided by the observer introduced at a later point will converge to the real state in at most $r_{min} + 1$ steps.

Recalling (2.34)–(2.35), one conclusion will be given about selecting an index T larger than the the minimal reconstructibility index r_{min} for reconstructibility analysis.

Theorem 3.12 *Assume that a BCN described by (2.34)–(2.35) is reconstructible. For any integer $\sigma \geq r_{min}$, each column of the matrix Ω_σ calculated by* (3.5) *contains at most one non-zero entry.*

Proof. The theorem is proven by induction. Let it start with $\sigma = r_{min}$. Since the BCN is reconstructible, according to Theorem 3.11 all columns of the matrix $\Omega_\sigma = \Omega_{r_{min}}$ contain at most one non-zero entry. Now assume that all columns of the matrix $\Omega_{\sigma-1}$ with $\sigma - 1 \geq r_{min}$ contain at most one non-zero entry. As mentioned before, each column of the matrix $\Omega_{\sigma-1}$ corresponds to the results calculated by different input and output trajectories represented by $y^*(0)u^*(0)y^*(1)u^*(1)\cdots y^*(\sigma-1) \in \Delta_{2^{(\sigma-1)\cdot(m+p)+p}}$. Two cases are investigated separately. In the first case, it is shown that if zero columns exist in $\Omega_{\sigma-1}$, then the corresponding columns in Ω_σ will also be zero columns. If an input and output trajectory represented by $y^*(0)u^*(0)y^*(1)u^*(1)\cdots y^*(\sigma-1)$ is not admissible, then the input and output trajectories represented by $y^*(0)u^*(0)y^*(1)u^*(1)\cdots y^*(\sigma-1)u^*(\sigma-1)y(\sigma)$ with any

$u(\sigma-1) \in \Delta_{2^m}$ and $y(\sigma) \in \Delta_{2^p}$ are still not admissible. That means the corresponding columns of the matrix Ω_σ are $\mathbf{0}_{2^n}$. Next, consider that if the columns in $\Omega_{\sigma-1}$ belonging to Δ_{2^n} exist, then the corresponding columns in Ω_σ belong to $\Delta_{2^n} \cup \{\mathbf{0}_{2^n}\}$. If an input and output trajectory represented by $y^*(0)u^*(0)y^*(1)u^*(1)\cdots y^*(\sigma-1)$ is admissible, then the corresponding column of the matrix $\Omega_{\sigma-1}$ is $\hat{x}^*(\sigma-1) \in \Delta_{2^n}$. Applying an input $u^*(\sigma-1) \in \Delta_{2^m}$ to the state $\hat{x}^*(\sigma-1)$ and according to (2.34)–(2.35), the output $y(\sigma) \in \Delta_{2^p}$ is generated and the state $\hat{x}(\sigma)$ at time σ is determined. The column of the matrix Ω_σ corresponding to the input and output trajectory with $y^*(\sigma) = y(\sigma)$ is equal to $\hat{x}(\sigma) \in \Delta_{2^n}$. If $y^*(\sigma) \neq y(\sigma)$, in conclusion, the input and output trajectory is inadmissible. Hence the corresponding columns of the matrix Ω_σ are $\mathbf{0}_{2^n}$. This proves that each column of the matrix Ω_σ contains at most one non-zero entry. Therefore, it follows from the inductive step that all columns of the matrix Ω_σ, $\sigma \geq r_{min}$ contain at most one non-zero entry. □

Remark 3.13 Theorem 3.12 shows that if the knowledge of the input and output trajectories of length $r_{min}+1$ can uniquely determine the final state, then the knowledge of the input and output trajectories of length $\sigma+1$ with $\sigma \geq r_{min}$ can also uniquely determine the final state. Hence, the minimal reconstructibility index r_{min} is an important index. Any value r larger than r_{min} can guarantee that all columns of the matrix $\Omega_{r_{min}}$ contains at most one non-zero entry, but the extra steps for reconstructibility analysis causes higher computational effort.

3.2.2 Recursive Method

Though the matrix Ω_r can be directly calculated according to (3.5), the index r for a BCN is usually unknown and should be determined. Naturally, exhaustive search can be applied to test each possible index $r \in \mathbb{Z}_+$ to check if the condition given in Theorem 3.11 is satisfied. However, this approach requires very high computational effort, especially if a BCN is not reconstructible. Hence, a recursive method will be proposed based on the following theorem.

Theorem 3.14 *The matrix Ω_r in* (3.5) *used in the explicit method can be recursively calculated as follows:*

$$\begin{aligned} \Omega_0 &= H^{\mathrm{T}}, \\ \Omega_j &= \Phi_{2^n}^{\mathrm{T}}\left(I_{2^n} \otimes H^{\mathrm{T}}\right) L \left(\Omega_{j-1} \otimes I_{2^{m+p}}\right),\ j = 1, 2, \cdots, r. \end{aligned} \tag{3.13}$$

Proof. The dimensions of the matrices are $H \in \mathbb{R}^{2^p \times 2^n}$, $I_{2^{m+n}} \otimes H \in \mathbb{R}^{2^{m+n+p} \times 2^{m+2n}}$ and $I_{2^{m+p}} \otimes H \in \mathbb{R}^{2^{m+2p} \times 2^{m+n+p}}$. Recall that the mixed-product property of Kronecker product is

$$(A \otimes B) \cdot (C \otimes D) = (A \cdot C) \otimes (B \cdot D). \tag{3.14}$$

Applying (3.14) and according to Definition 2.1,

$$\begin{aligned}
H\left(I_{2^{m+n}} \otimes H\right) &= \left(H \otimes I_{2^{m+p}}\right) \cdot \left(I_{2^{m+n}} \otimes H\right) \\
&= \left(\underbrace{H}_{A} \otimes I_{2^{m+p}}\right) \cdot \left(I_{2^n} \otimes \underbrace{\left(I_{2^m} \otimes H\right)}_{B}\right) \\
&= \left(I_{2^p} \otimes \underbrace{\left(I_{2^m} \otimes H\right)}_{B}\right) \cdot \left(\underbrace{H}_{A} \otimes I_{2^{m+n}}\right) \\
&= \left(I_{2^{m+p}} \otimes H\right) \cdot \left(H \otimes I_{2^{m+n}}\right) = \left(I_{2^{m+p}} \otimes H\right) H
\end{aligned} \tag{3.15}$$

is obtained. Let a matrix be $\tilde{M}_k \in \mathbb{R}^{2^{(k-1)(m+p)+p} \times 2^{(k-1)(m+n)+n}}$, $\forall k \in \mathbb{N}^+$. Applying (3.14), a similar result can be obtained that

$$\begin{aligned}
\tilde{M}_k\left(I_{2^{k(m+n)}} \otimes H\right) &= \left(\tilde{M}_k \otimes I_{2^{m+p}}\right) \cdot \left(I_{2^{k\cdot(m+n)}} \otimes H\right) \\
&= \left(\underbrace{\tilde{M}_k}_{A} \otimes I_{2^{m+p}}\right) \cdot \left(I_{2^{(k-1)\cdot(m+n)+n}} \otimes \underbrace{\left(I_{2^m} \otimes H\right)}_{B}\right) \\
&= \left(I_{2^{(k-1)\cdot(m+p)+p}} \otimes \underbrace{\left(I_{2^m} \otimes H\right)}_{B}\right) \cdot \left(\underbrace{\tilde{M}_k}_{A} \otimes I_{2^{m+n}}\right) \\
&= \left(I_{2^{k\cdot(m+p)}} \otimes H\right) \cdot \left(\tilde{M}_k \otimes I_{2^{m+n}}\right) \\
&= \left(I_{2^{k\cdot(m+p)}} \otimes H\right) \tilde{M}_k.
\end{aligned} \tag{3.16}$$

Because $H(I_{2^{m+n}} \otimes H)$ has the dimension $2^{m+2p} \times 2^{m+2n}$, the following result can be got by using (3.16):

$$H\left(I_{2^{m+n}} \otimes H\right)\left(I_{2^{2(m+n)}} \otimes H\right) = \left(I_{2^{2(m+p)}} \otimes H\right)\left(I_{2^{m+p}} \otimes H\right) H \in \mathbb{R}^{2^{2(m+p)+p} \times 2^{2(m+n)+n}}.$$

Inductive reasoning by using (3.16) shows that

$$\prod_{i=0}^{r}(I_{2^{i\cdot(n+m)}} \otimes H) = \prod_{j=r}^{0}(I_{2^{j\cdot(p+m)}} \otimes H). \tag{3.17}$$

Based on (3.17) and recalling (2.2) in Proposition 2.2, (3.5) can be equivalently rewritten as

$$\begin{aligned} \Omega_r &= (\Phi_{2^n}^{\mathrm{T}} L_{eq})^r \left(\prod_{i=0}^{r}(I_{2^{i\cdot(n+m)}} \otimes H)\right)^{\mathrm{T}} \\ &= (\Phi_{2^n}^{\mathrm{T}} L_{eq})^r \left(\prod_{i=r}^{0}(I_{2^{i\cdot(p+m)}} \otimes H)\right)^{\mathrm{T}} \\ &= (\Phi_{2^n}^{\mathrm{T}} L_{eq})^r \prod_{i=0}^{r}(I_{2^{i\cdot(p+m)}} \otimes H^{\mathrm{T}}) \end{aligned} \tag{3.18}$$

which shows that a recursive form can be evaluated to calculate the matrix Ω_r as follows:

$$\begin{aligned} &\Omega_0 = H^{\mathrm{T}}, \\ &\Omega_j = \Phi_{2^n}^{\mathrm{T}} L_{eq} \Omega_{j-1}\left(I_{2^{j\cdot(m+p)}} \otimes H^{\mathrm{T}}\right), \, j = 1, 2, \cdots, r. \end{aligned} \tag{3.19}$$

According to (3.12), the matrix Ω_{j-1} has the dimension $2^n \times 2^{(j-1)\cdot(m+p)+p}$. Because $\left(I_{2^{j\cdot(m+p)}} \otimes H^{\mathrm{T}}\right)$ has the dimension $2^{j\cdot(m+p)+n} \times 2^{j\cdot(m+p)+p}$, the following equation can be obtained by using mixed-product property of the Kronecker product:

$$\begin{aligned} \Omega_{j-1}\left(I_{2^{j(m+p)}} \otimes H^{\mathrm{T}}\right) &= (\Omega_{j-1} \otimes I_{2^{n+m}}) \cdot (I_{2^{j\cdot(m+p)}} \otimes H^{\mathrm{T}}) \\ &= \left(\Omega_{j-1} \otimes I_{2^{n+m}}\right) \cdot \left(I_{2^{(j-1)\cdot(m+p)+p}} \otimes I_{2^m} \otimes H^{\mathrm{T}}\right) \\ &= \left(I_{2^n} \otimes \underbrace{I_{2^m} \otimes H^{\mathrm{T}}}_{B}\right) \cdot \left(\underbrace{\Omega_{j-1}}_{A} \otimes I_{2^{m+p}}\right) \\ &= (I_{2^{n+m}} \otimes H^{\mathrm{T}}) \cdot (\Omega_{j-1} \otimes I_{2^{m+p}}) \\ &= (I_{2^{n+m}} \otimes H^{\mathrm{T}})(\Omega_{j-1} \otimes I_{2^{m+p}}). \end{aligned}$$

Similarly, as $L_{eq} \in \mathbb{R}^{2^n \times 2^{m+n}}$ and $I_{2^{n+m}} \otimes H^{\mathrm{T}} \in \mathbb{R}^{2^{2n+m} \times 2^{n+m+p}}$, there is

$$L_{eq}(I_{2^{n+m}} \otimes H^{\mathrm{T}}) = (I_{2^n} \otimes H^{\mathrm{T}})L_{eq}. \tag{3.20}$$

With the results obtained above, it has

$$\Omega_j = \Phi_n^{\mathrm{T}}(I_{2^n} \otimes H^{\mathrm{T}})L_{eq}(\Omega_{j-1} \otimes I_{2^{m+p}}) \tag{3.21}$$

which shows (3.19) is equivalent to (3.13). □

It is worth mentioning that the reason for using the recursive form (3.13) is to introduce an efficient approach for reconstructibility analysis. In addition, the columns of the matrix Ω_{j-1} can be easily deleted without adapting the dimension according to definition of STP, i.e. Definition 2.1. Based on (3.13), a recursive method requiring lower computational effort is introduced.

Theorem 3.15 *Given a BCN described by (2.34)–(2.35). Initialize $\Omega_0 = H^{\mathrm{T}}$ and calculate the matrices Ω_j, $j = 1, 2, \cdots$ forward as*

$$\Omega_j = \Phi_{2^n}^{\mathrm{T}} \left(I_{2^n} \otimes H^{\mathrm{T}}\right) L_{eq} \left(\Theta_{j-1} \otimes I_{2^{m+p}}\right) \tag{3.22}$$

where Θ_{j-1} is obtained from the matrix Ω_{j-1} by deleting the duplicate columns and those columns with at most one non-zero entry. The BCN (2.34)–(2.35) is reconstructible if and only if there is a non-negative integer r, so that all columns in the matrix Ω_j contain at most one non-zero entry.

Proof. Recall that the calculation of the matrix Ω_r according to (3.5) can be simplified by using the recursive form (3.13). For reconstructibility analysis, only the admissible input and output trajectories are of interest. Therefore, only the non-zero columns in Ω_{j-1} need to be considered. If the final state can be uniquely determined by the knowledge of some admissible input and output trajectories (i.e. the columns in Ω_{j-1} contain only one non-zero entry), then given any possible input and according to (2.32), the successor state is also uniquely determined. Furthermore, assume that there are two duplicate columns in the matrix Ω_{j-1}, i.e. $Col_i(\Omega_{j-1})$ and $Col_k(\Omega_{j-1}), i \neq k$, so that $Col_i(\Omega_{j-1}) = Col_k(\Omega_{j-1})$. In this case, one has $\Phi_{2^n}^{\mathrm{T}} \left(I_{2^n} \otimes H^{\mathrm{T}}\right) L_{eq} Col_i(\Omega_{j-1}) = \Phi_{2^n}^{\mathrm{T}} \left(I_{2^n} \otimes H^{\mathrm{T}}\right) L_{eq} Col_k(\Omega_{j-1})$, which shows that it is sufficient to consider only $Col_i(\Omega_{j-1})$ or $Col_k(\Omega_{j-1})$ to check reconstructibility.

As the matrix Ω_j has a large number of rows, the multiplication between the matrix $\Phi_{2^n}^{\mathrm{T}}(I_{2^n} \otimes H^{\mathrm{T}})L_{eq}$ and one column in the matrix Ω_j requires much higher computational effort than finding and deleting duplicate columns. In order to reduce computational time and space complexity, the duplicate columns and the columns with at most one non-zero entry are deleted from Ω_j before Ω_{j+1} is calculated. Based on the analysis above and Theorem 3.11, the following result is obtained. □

Remark 3.16 If there is a positive integer $k < r_{min}$ so that some columns of the matrix Ω_k calculated according to (3.22) contains only one non-zero entry, then this implies that a state observer may provide the correct state estimate earlier than the r_{min} steps.

It is important to note that for a reconstructible BCN the recursive calculation of Ω_j will be stopped after a maximum of r_{min} steps. However, if a BCN is not reconstructible, then the procedure to calculate Ω_j will not terminate in finite time. In order to solve this problem, an additional termination condition described below can be used to stop the calculation of Ω_j.

Theorem 3.17 *Consider a BCN described by (2.34)–(2.35) and assume that* $\Omega_0 = H^{\mathrm{T}}$ *and let* Ω_j, $j = 1, 2, \cdots$ *be calculated recursively by* (3.22). *The BCN is not reconstructible, if and only if there exists a non-negative integer* σ, *so that*

$$Col(\Theta_{\sigma+1}) = Col(\Theta_\sigma) \neq \emptyset. \tag{3.23}$$

Proof. (Sufficiency) Assume that there is a non-negative integer σ so that (3.23) holds. This implies $Col(\Theta_{t+1}) = Col(\Theta_t)$, $\forall t \geq \sigma$. Let a matrix Θ be defined so that $Col(\Theta) = Col(\Theta_t)$, $\forall t \geq \sigma$. According to Theorem 3.15, all columns of the matrix $L\Phi_{2^n}^{\mathrm{T}}(I_{2^n} \otimes H_{new}^{\mathrm{T}})Col_k(\Theta)$, $\forall k$ either are deleted or belong to the set $Col(\Theta)$. As each entry of the matrix Θ is either 1 or 0, the set $Col(\Theta)$ always has finite number of column vectors. There are $S \in \mathbb{Z}_+$ positive integers $k_1, k_2, \cdots, k_S$ and indices $i_1, i_2, \cdots, i_S \in \{1, 2, \cdots, 2^{p+m}\}$ such that

$$\begin{aligned}
\Phi_{2^n}^{\mathrm{T}}\left(I_{2^n} \otimes H^{\mathrm{T}}\right) L_{eq} Col_{k_1}(\Theta)\delta_{2^{p+m}}^{i_1} &= Col_{k_2}(\Theta),\\
\Phi_{2^n}^{\mathrm{T}}\left(I_{2^n} \otimes H^{\mathrm{T}}\right) L_{eq} Col_{k_2}(\Theta)\delta_{2^{p+m}}^{i_2} &= Col_{k_3}(\Theta),\\
&\vdots\\
\Phi_{2^n}^{\mathrm{T}}\left(I_{2^n} \otimes H^{\mathrm{T}}\right) L_{eq} Col_{k_S}(\Theta)\delta_{2^{p+m}}^{i_S} &= Col_{k_1}(\Theta).
\end{aligned} \tag{3.24}$$

As all columns $Col_{k_1}(\Theta), Col_{k_2}(\Theta), \cdots, Col_{k_S}(\Theta)$ contain more than one non-zero entry, (3.24) tells us that the knowledge of the input and output trajectory $\delta_{2^{p+m}}^{i_1}, \delta_{2^{p+m}}^{i_2}, \cdots, \delta_{2^{p+m}}^{i_S}$ of period S can not determine the final state uniquely. Hence, the BCN is not reconstructible.

(Necessity) Suppose that by contradiction, (3.23) does not hold (i.e. $Col(\Theta_{t+1}) \neq Col(\Theta_t), \forall t \in \mathbb{N}$) and the BCN is reconstructible. Recall that L_{eq} is a logical matrix, i.e. each column of the matrix L_{eq} contains one non-zero entry 1. The matrix $\Phi_{2^n}^{\mathrm{T}}\left(I_{2^n} \otimes H^{\mathrm{T}}\right) L_{eq}$ can be split into 2^{n+m} equal blocks. The i-th block can be expressed as $\Phi_{2^n}^{\mathrm{T}}\left(I_{2^n} \otimes H^{\mathrm{T}}\right) Col_i(L_{eq})$. According to Proposition 2.5, one has

$$\begin{aligned}\Phi_{2^n}^{\mathrm{T}}\left(I_{2^n} \otimes H^{\mathrm{T}}\right) Col_i(L_{eq}) &= \Phi_{2^n}^{\mathrm{T}} Col_i(L_{eq}) H^{\mathrm{T}} \\ &= \Phi_{2^n}^{\mathrm{T}} Col_i(L_{eq})[Col_1(H^{\mathrm{T}})\ Col_2(H^{\mathrm{T}})\ \cdots\ Col_{2^n}(H^{\mathrm{T}})].\end{aligned}$$

Recall that the matrix H is a logical matrix. For any column $Col_i(L_{eq})$, there is only one column $Col_j(H^{\mathrm{T}})$, such that $Col_i(L_{eq}) \odot Col_j(H^{\mathrm{T}}) \neq \mathbf{0}_{2^n}$. Therefore, the number of the non-zero entry 1 in the vector $Col_{k_t}(\Omega_t) = \Phi_{2^n}^{\mathrm{T}}\left(I_{2^n} \otimes H^{\mathrm{T}}\right) L_{eq}$ $Col_{k_{t-1}}(\Omega_{t-1})\delta_{2^{p+m}}^{i_{t-1}}, \forall t \in \mathbb{Z}_+, i_{t-1} = 1, 2, \cdots, 2^{m+p}$ is not more than the number of the non-zero entry 1 in the column vector $Col_{k_{t-1}}(\Omega_{t-1})$. If an integer $\sigma \in \mathbb{N}$ exists so that $Col_{k_t}(\Omega_t)$ and $Col_{k_{t-1}}(\Omega_{t-1}), \forall t > \sigma$ have the same number of the non-zero entry 1, then an integer $\lambda \in \mathbb{Z}_+$ can always be found so that the vectors $Col_{k_\sigma}(\Omega_\sigma), Col_{k_{\sigma+1}}(\Omega_{\sigma+1}), \cdots, Col_{k_{\sigma+\lambda}}(\Omega_{\sigma+\lambda})$ are in a cycle. The input and output trajectory corresponding to the cycle is $\{y(\sigma), u(\sigma), y(\sigma+1), u(\sigma+1), \cdots, y(\sigma+\lambda), u(\sigma+\lambda)\}$. According to Theorem 4 in Fornasini and Valcher (2013), in this case the BCN is not reconstructible, which contradicts the assumption. Otherwise, if the number of 1 in $Col_{k_t}(\Omega_t)$ is smaller than $Col_{k_{t-1}}(\Omega_{t-1})$ (i.e. $\mathbf{1}_{2^n}^{\mathrm{T}} Col_{k_t}(\Omega_t) < \mathbf{1}_{2^n}^{\mathrm{T}} Col_{k_{t-1}}(\Omega_{t-1})$), then there is an integer $\sigma \in \mathbb{Z}_+$ so that $Col_{k_{t+\sigma}}(\Omega_{t+\sigma}) \in \Delta_{2^n} \cup \{\mathbf{0}_{2^n}\}$. Hence they are deleted from the matrix Ω_σ. So, it can be seen that $Col(\Theta_t) \neq Col(\Theta_{t-1}), \forall t \in \mathbb{Z}_+$. Furthermore, if there is an integer σ so that all columns of the matrix Θ_σ are deleted, then according to Theorem 3.15, the BCN is reconstructible and the assumption would be contradicted. □

If the condition (3.23) is satisfied, then the procedure to calculate Ω_j according to (3.22) is terminated and it can be concluded that the BCN is not reconstructible. In this case, the state estimate provided by an observer (including the Luenberger-like observer designed in Section 4.1) cannot be guaranteed to always converge to the real state.

Note that Fornasini and Valcher (2013) pointed out that if the knowledge of the input and output trajectories of length $r + 1$ can uniquely determine the final state

Algorithm 1: Given the BCN (2.32)–(2.33). Check reconstructibility of the BCN.

1. Initialize $j = 0$.
2. Compute the matrix $\Omega_j = H^{\mathrm{T}}$.
3. Delete the duplicate columns of Ω_j and the columns of Ω_j which have at most one non-zero entry. The result is denoted as Θ_j.
4. Calculate the matrix Ω_{j+1} according to (3.22).
5. If all columns of the matrix Ω_{j+1} contain at most one non-zero entry, then stop. The BCN is reconstructible. Otherwise, if $Col(\Theta_{j+1}) = Col(\Theta_j)$, then calculation of the matrix Ω_j according to (3.22) is stopped, the BCN is not reconstructible and the state estimate provided by an observer cannot always converge to the real state. If neither is the case, let $j = j + 1$ and return to Step 4.

(i.e. BCN is reconstructible), then r has the upper bound $(2^n + 1) \cdot (2^n - 2)/2$. For convenience, the above result is summarized in Algorithm 1.

3.2.3 Example

In order to illustrate the result for reconstructibility analysis, a simple BCN is considered, which is a reduced Boolean model for the *lac* operon in the bacterium *Escherichia coli* derived in Veliz-Cuba and Stigler (2011) and also used in Li et al. (2013). Assume that the genes X_1 and X_2 can be measured, X_3 can not be measured. Then the BCN can be described by

$$\begin{cases} X_1(t+1) = \neg U_1(t) \wedge (X_2(t) \vee X_3(t)), \\ X_2(t+1) = \neg U_1(t) \wedge U_2(t) \wedge X_1(t), \\ X_3(t+1) = \neg U_1(t) \wedge (U_2(t) \vee (U_3(t) \wedge X_1(t))). \end{cases} \tag{3.25}$$

In the model, X_1 represents the transcription of mRNA, the states X_2 and X_3 indicate, respectively, a high and a medium concentration of lactose, U_1 represents an abundance of extra cellular glucose, U_2 and U_3 denote, respectively, a high and a medium concentration of extra cellular lactose. Suppose that two outputs $Y_1(t)$ and $Y_2(t)$ are measured, i.e.

$$\begin{cases} Y_1(t) = X_1(t), \\ Y_2(t) = X_2(t). \end{cases} \tag{3.26}$$

Using STP, (3.25)–(3.26) can be converted into the same form as (2.34) and (2.35) with $x(t) = \ltimes_{i=1}^{3} x_i(t)$, $u(t) = \ltimes_{i=1}^{3} u_i(t)$, $y(t) = \ltimes_{i=1}^{2} y_i(t)$ and

$$\begin{aligned} L_{eq} = \delta_8[&8\ 8\ 8\ 8\ 1\ 1\ 3\ 4\ 8\ 8\ 8\ 8\ 1\ 1\ 3\ 4\ 8\ 8\ 8\ 8\ 1\ 1\ 3\ 4\ 8\ 8\ 8\ 8\ 5\ 5\ 7\ 8\\ &8\ 8\ 8\ 8\ 3\ 3\ 4\ 4\ 8\ 8\ 8\ 8\ 3\ 3\ 4\ 4\ 8\ 8\ 8\ 8\ 3\ 3\ 4\ 4\ 8\ 8\ 8\ 8\ 7\ 7\ 8\ 8],\\ H = \delta_4[&1\ 1\ 2\ 2\ 3\ 3\ 4\ 4]. \end{aligned}$$

Now the reconstructibility of the BCN (3.25)–(3.26) will be investigated. In order to apply the explicit method, initialize the index $r = 1$. The matrix $\Omega_1 \in \mathbb{R}^{8\times 128}$ is calculated according to (3.5). It is found out that all the columns of Ω_1 are one of the following column vectors:

$$\begin{bmatrix}0\\0\\0\\0\\0\\0\\0\\0\end{bmatrix}, \begin{bmatrix}0\\0\\0\\0\\0\\0\\0\\1\end{bmatrix}, \begin{bmatrix}0\\0\\0\\0\\0\\0\\1\\0\end{bmatrix}, \begin{bmatrix}0\\0\\0\\0\\1\\0\\0\\0\end{bmatrix}, \begin{bmatrix}0\\0\\0\\1\\0\\0\\0\\0\end{bmatrix}, \begin{bmatrix}0\\0\\1\\0\\0\\0\\0\\0\end{bmatrix}, \begin{bmatrix}1\\0\\0\\0\\0\\0\\0\\0\end{bmatrix}. \tag{3.27}$$

It is straightforward to see that the matrix Ω_1 only contains columns with less than two non-zero entries. According to Theorem 3.11, the BCN (3.25)–(3.26) is reconstructible.

Next, the recursive method is applied. According to (3.13), the matrix Ω_0 is initialized as H^{T}. The following column vectors can be found in the matrix Ω_0:

$$\begin{bmatrix}1\\1\\0\\0\\0\\0\\0\\0\end{bmatrix}, \begin{bmatrix}0\\0\\1\\1\\0\\0\\0\\0\end{bmatrix}, \begin{bmatrix}0\\0\\0\\0\\1\\1\\0\\0\end{bmatrix}, \begin{bmatrix}0\\0\\0\\0\\0\\0\\1\\1\end{bmatrix}.$$

As there is neither zero columns nor the columns with one non-zero entry in the matrix Ω_0, the matrix Θ_0 is equal to the matrix Ω_0, i.e., $\Theta_0 = \Omega_0$. Then the matrix Ω_1 is calculated. It can be found out that all the columns of the matrix Ω_1 are one of the column vectors in (3.27). According to Theorem 3.15, the BCN (3.25)–(3.26) is therefore reconstructible and the procedure is stopped. The minimal reconstructibility index is $r_{min} = 1$.

In this example, it can be shown that observability and reconstructibility of BCNs are not equivalent. For example, with the knowledge of any input and output trajectory $\{(y(t), u(t)), t = 0, 1, \cdots, r\}, r \in \mathbb{N}$ with the input $u(0) \in \{\delta_8^3, \delta_8^4, \delta_8^7, \delta_8^8\}$ and output $y(0) = \delta_4^1$, all the possible initial states at time $t = 0$ are $\{\delta_8^1, \delta_8^2\}$. Applying the input $u(0)$, the possible state at the next time is $\{\delta_8^8\}$. Hence, there is no chance to uniquely determine the initial state $x(0)$. According to Definition 3.4, the BCN (3.25)–(3.26) is not observable.

Now assume that only output Y_2 is measured. In this case, the logical matrix H is

$$H = \delta_2[1\ 1\ 2\ 2\ 1\ 1\ 2\ 2]. \tag{3.28}$$

Following the same procedure, at first the matrix Ω_0 is initialized as $\Omega_0 = H^{\mathrm{T}}$. As each column of the matrix Ω_0 contains more than one non-zero entry, no columns will be removed, i.e. the matrix Θ_0 is set to Ω_0. Based on this, the matrix Ω_1 is calculated according to (3.13). The columns equal to $\mathbf{0}_8$, δ_8^1, δ_8^3, δ_8^4 and δ_8^8 contain at most one non-zero entry and are therefore removed. The matrix Θ_1 is composed by the column vectors

$$\begin{bmatrix}0\\0\\0\\1\\0\\0\\0\\1\end{bmatrix}, \begin{bmatrix}0\\0\\1\\0\\0\\0\\1\\0\end{bmatrix}, \begin{bmatrix}0\\0\\1\\1\\0\\0\\0\\0\end{bmatrix}, \begin{bmatrix}0\\0\\1\\1\\0\\0\\1\\1\end{bmatrix}, \begin{bmatrix}1\\0\\0\\0\\1\\0\\0\\0\end{bmatrix}. \tag{3.29}$$

Then the matrices Θ_2 and Θ_3 are calculated. It can be found out that all the columns of Θ_1 and Θ_2 are the same and are one of the column vectors

$$\begin{bmatrix}0\\0\\0\\0\\0\\0\\1\\1\end{bmatrix},\begin{bmatrix}0\\0\\0\\1\\0\\0\\0\\1\end{bmatrix},\begin{bmatrix}0\\0\\1\\0\\0\\0\\1\\0\end{bmatrix},\begin{bmatrix}0\\0\\1\\1\\0\\0\\0\\0\end{bmatrix},\begin{bmatrix}0\\0\\1\\1\\0\\0\\1\\1\end{bmatrix},\begin{bmatrix}1\\0\\0\\0\\1\\0\\0\\0\end{bmatrix}. \tag{3.30}$$

This means that the stopping criteria (3.23) holds for $\sigma = 2$. Based on Theorem 3.17, the procedure to calculate Ω_j is terminated and it can be concluded that the BCN (2.32)–(2.33) is not reconstructible. In this example, by making use of the termination condition in Theorem 6, the recursive method provides the result of reconstructibility analysis after only two steps. If the calculation of the matrix Ω_j according to (3.13) is terminated after reaching the upper bound of recursion steps given in Fornasini and Valcher (2013), then the recursive algorithm needs to run $(2^3 + 1) \cdot (2^3 - 2)/2 = 27$ steps, which is much larger than the case of applying the termination condition.

3.3 Reconstructibility for Boolean Control Networks With Unknown Inputs

The approaches for reconstructibility analysis of BCNs given in Section 3.2 can not take unknown inputs, such as disturbances and noises, into account. In this section, disturbances and noises as unknown inputs are considered in the model and the unknown input decouplability of BCNs is investigated. The basic idea is to take all possible unknown inputs into account with the help of the all-one column vector.

3.3.1 Boolean Control Networks With Unknown Inputs

A class of BCNs with unknown inputs is introduced, which can be described as follows:

$$x(t+1) = L\xi(t)x(t)u(t), \tag{3.31}$$

$$y(t) = H\omega(t)x(t) \tag{3.32}$$

where $x(t) \in \Delta_{2^n}$, $u(t) \in \Delta_{2^m}$, $y(t) \in \Delta_{2^p}$ are, respectively, the states, the known inputs and the measured outputs, $\xi(t) \in \Delta_{2^b}$ and $\omega(t) \in \Delta_{2^q}$ are the unknown inputs. To demonstrate the influence of unknown inputs ξ and ω, two examples are given in the following.

Consider the case that $\xi_1, \xi_2, \cdots, \xi_b$ can only influence the known input $U_1(t), U_2(t), \cdots, U_m(t)$. This relationship can be described by the following boolean equations:

$$\begin{cases} \tilde{U}_1(t) = g_{U_1}(U_1(t), U_2(t), \cdots, U_m(t), \xi_1(t), \xi_2(t), \cdots, \xi_b(t)), \\ \tilde{U}_2(t) = g_{U_2}(U_1(t), U_2(t), \cdots, U_m(t), \xi_1(t), \xi_2(t), \cdots, \xi_b(t)), \\ \quad\vdots \\ \tilde{U}_m(t) = g_{U_m}(u_1(t), U_2(t), \cdots, U_m(t), \xi_1(t), \xi_2(t), \cdots, \xi_b(t)) \end{cases} \tag{3.33}$$

where $\tilde{U}_1(t), \tilde{U}_2(t), \cdots, \tilde{U}_m(t)$, $U_1(t), U_2(t), \cdots, U_m(t)$ and $\xi_1, \xi_2, \cdots, \xi_b$ are, respectively, the resulting inputs, the known inputs and the unknown inputs, $g_{U_1}, g_{U_2}, \cdots, g_{U_m}$ are the logical functions that describe the influence of unknown inputs. According to Cheng and Qi (2010), (3.33) can be written in an algebraic form as

$$\tilde{u}(t) = M_u \xi(t) u(t) \tag{3.34}$$

where M_u is the logical matrix corresponding to the Boolean functions g_{U_1}, $g_{U_2}, \cdots, g_{U_m}$. In order to consider the resulting inputs, (2.32) can be modified as

$$x(t+1) = L_u x(t) \tilde{u}(t) \text{ with } L_{\tilde{u}} \in \mathcal{L}_{2^n \times 2^{m+n}}. \tag{3.35}$$

Replacing the variable $\tilde{u}$ with (3.34), (3.35) can be written as

$$\begin{aligned} x(t+1) &= L_u x(t) M_u \xi(t) u(t) \\ &= L_u (I_{2^n} \otimes M_u) x(t) \xi(t) u(t) \\ &= L_u (I_{2^n} \otimes M_u) W_{[2^b, 2^n]} \xi(t) x(t) u(t). \end{aligned} \tag{3.36}$$

Let the matrix L be replaced by $L_u (I_{2^n} \otimes M_u) W_{[2^b, 2^n]}$. It can be seen that (3.36) has the same algebraic form as (3.31).

The influence of $\omega_1, \omega_2, \cdots, \omega_q$ on the output can be described by the following Boolean equations:

$$\begin{cases} Y_1(t) &= g_{Y_1}(\tilde{Y}_1(t), \tilde{Y}_2(t), \cdots, \tilde{Y}_p(t), \omega_1(t), \cdots, \omega_q(t)), \\ Y_2(t) &= g_{Y_2}(\tilde{Y}_1(t), \tilde{Y}_2(t), \cdots, \tilde{Y}_p(t), \omega_1(t), \cdots, \omega_q(t)), \\ &\vdots \\ Y_p(t) &= g_{Y_p}(\tilde{Y}_1(t), \tilde{Y}_2(t), \cdots, \tilde{Y}_p(t), \omega_1(t), \cdots, \omega_q(t)), \end{cases} \tag{3.37}$$

where $Y_1(t), Y_2(t), \cdots, Y_p(t)$, $\tilde{Y}_1(t), \tilde{Y}_2(t), \cdots, \tilde{Y}_p(t)$ and $\omega_1, \omega_2, \cdots, \omega_q$ are, respectively, the measured outputs, the real outputs. The unknown inputs, $g_{Y,1}$, $g_{Y,2}, \cdots, g_{Y,p}$ are the logical functions that describe the effect of the unknown inputs. According to Cheng and Qi (2010), (3.37) can be written in an algebraic form as

$$y(t) = M_y \omega(t) \tilde{y}(t). \tag{3.38}$$

Based on (2.33), it is clear that $\tilde{y}(t) = H_y x(t)$. Then (3.38) can be further written as

$$\begin{aligned} y(t) =& M_y \omega(t) H_y x(t) \\ =& M_y (I_{2^q} \otimes H_y) \omega(t) x(t). \end{aligned}$$

Considering $M_y(I_{2^q} \otimes H_y)$ as the matrix H, the output can be written in the form of (3.32).

3.3.2 Decoupling the Effect of Unknown Input

The main purpose of this section is to analyze the condition under which it is possible to build an observer to provide the exact estimation of the state of a BCN with unknown inputs described by (3.31)–(3.32). At first, the definition of unknown input decouplability will be introduced. After that, a necessary and sufficient condition will be given.

Definition 3.18 (Unknown Input Decouplability) Consider the BCN with unknown inputs (3.31)–(3.32). The unknown inputs $\xi(t) \in \Delta_{2^b}$ and $\omega(t) \in \Delta_{2^q}$ are said to be decouplable if and only if for any possible unknown inputs $\xi(t)$ and $\omega(t)$ a non-negative integer r can be found so that with the knowledge of every admissible input and output trajectory $\{y(0), u(0), \cdots, y(t), u(t)\}, t \geq r$, the final state $x(t)$ can be correctly determined.

Note that the unknown input decouplability in Definition 3.18 implies that the unknown inputs should have no influence on the state estimation error generated by an observer.

In the following, an approach to check the unknown input decouplability defined in Definition 3.18 for the BCN with unknown inputs (3.31)–(3.32) is proposed. Assume that the state $x(t)$ and the input $u(t)$ are known. According to (3.31) and (3.32), the state $x(t+1)$ and the output $y(t+1)$ under the influence of the unknown input $\xi(t) = \delta_{2^b}^i$ and $\omega(t+1) = \delta_{2^q}^j$ are determined by

$$x(t+1) = L\delta_{2^b}^i x(t)u(t), \tag{3.39}$$

$$y(t+1) = H\delta_{2^q}^j x(t+1) = H\delta_{2^q}^j L\delta_{2^b}^i x(t)u(t). \tag{3.40}$$

To consider all possible effects of the unknown input $\xi(t)$, $x(t+1)$ is calculated for the unknown input $\xi(t) = \delta_{2^b}^i$, $i = 1, 2, \cdots, 2^b$ by (3.39) and summed up. This results in

$$\begin{aligned}\bar{x}(t+1) =& \sum_{i=1}^{2^b} L\delta_{2^b}^i x(t)u(t)\\ =& L\left(\sum_{i=1}^{2^b} \delta_{2^b}^i\right) x(t)u(t)\\ =& L\mathbf{1}_{2^b} x(t)u(t)\end{aligned} \tag{3.41}$$

where $\bar{x}(t+1)$ is a vector of dimension $2^n \times 1$ and $[\bar{x}(t+1)]_k \neq 0$ means that there is at least one possible unknown input $\xi(t) \in \Delta_{2^b}$, so that with the unknown input $\xi(t)$ and applying the input $u(t)$ the state $x(t)$ is transient to the state $\delta_{2^n}^k$ at time $t+1$.

Furthermore, assume that the output $y(t+1)$ is known. For an unknown input $\omega(t+1) = \delta_{2^q}^j$, states that can generate the output $y(t+1)$ can be determined by

$$\begin{aligned}\bar{x}_{o,j}(t+1) =& (H\delta_{2^q}^j)^{\mathrm{T}} y(t+1)\\ =& (\delta_{2^q}^j)^{\mathrm{T}} H^{\mathrm{T}} y(t+1)\end{aligned} \tag{3.42}$$

where $\delta_{2^n}^k$ is a candidate of the internal state x, only if $[\bar{x}_{o,j}(t+1)]_k \neq 0$. Taking all possible effects of the unknown input $\omega(t+1) \in \Delta_{2^q}$ into consideration, states that can possibly generate the output $y(t+1)$ are determined by

$$\begin{aligned}\bar{x}_o(t+1) =& \sum_{j=1}^{2^q} \bar{x}_{o,j}(t+1) \\ =& (H\mathbf{1}_{2^q})^{\mathrm{T}} y(t+1).\end{aligned} \tag{3.43}$$

If $\exists j \in \{1, 2, \cdots, 2^q\}$ so that $[\bar{x}_{o,j}(t+1)]_k \neq 0$, then one obtains $[\bar{x}_o(t+1)]_k \neq 0$. In order to find states comprised in vector $\bar{x}(t+1)$ that can possibly generate the output $y(t+1)$, the element-wise multiplication (2.11) is applied. That is

$$\begin{aligned}\bar{x}(t+1) \odot \bar{x}_o(t+1) =& \Phi_{2^n}^{\mathrm{T}} (I_{2^n} \otimes (H\mathbf{1}_{2^q})^{\mathrm{T}}) L\mathbf{1}_{2^b} x(t)u(t)y(t+1) \\ =& \Phi_{2^n}^{\mathrm{T}} (I_{2^n} \otimes (H \sum_{\substack{1\leq j\leq 2^q, \\ j\neq j^*}} \delta_{2^q}^j)^{\mathrm{T}}) L (\sum_{\substack{1\leq i\leq 2^b, \\ i\neq i^*}} \delta_{2^b}^i) x(t)u(t)y(t+1) \\ &+ \Phi_{2^n}^{\mathrm{T}} \left(I_{2^n} \otimes \left(H\delta_{2^q}^{j^*}\right)^{\mathrm{T}}\right) L \left(\delta_{2^b}^{i^*}\right) x(t)u(t)y(t+1)\end{aligned} \tag{3.44}$$

which shows that if there are unknown inputs $\xi(t) = \delta_{2^b}^{i^*}$ and $\omega(t+1) = \delta_{2^q}^{j^*}$, so that the state $x(t+1) = \delta_{2^n}^k$ can be reached from the state $x(t)$ and generates the output $y(t+1)$, then $[\bar{x}(t+1) \odot \bar{x}_o(t+1)]_k \neq 0$. As all the vectors $\Phi_{2^n}^{\mathrm{T}} \left(I_{2^n} \otimes \left(H\delta_{2^q}^j\right)^{\mathrm{T}}\right) L \left(\delta_{2^b}^i\right), \forall i, j$ contain only non-negative entries, if $\bar{x}(t+1) \odot \bar{x}_o(t+1) \in \Delta_{2^n}$, then the state at time $t+1$ can be correctly estimated, no matter what unknown inputs $\xi(t)$ and $\omega(t+1)$ are. This means that state estimation is decoupled from unknown inputs. Therefore, to achieve decouplability of unknown inputs ξ and ω, a non-negative integer r should be found so that $\bar{x}(t+1) \odot \bar{x}_o(t+1) \in \Delta_{2^n}$ is satisfied for $t \geq r$.

From the analysis above, the matrices $L\mathbf{1}_{2^b}$ and $H\mathbf{1}_{2^q}$ contain all effects of the unknown inputs ξ and ω. Let

$$\tilde{L} = L\mathbf{1}_{2^b}, \ \tilde{H} = H\mathbf{1}_{2^q}. \tag{3.45}$$

Motivated by Theorem 3.15, the following conclusion can be obtained.

Theorem 3.19 *State estimation of the BCN with unknown inputs* (3.31)–(3.32) *can be decoupled from the unknown inputs ξ and ω if and only if the matrices $\tilde{L}$ and $\tilde{H}$ satisfy the following two conditions:*

(a) Let $\Omega_0 = \tilde{H}^{\mathrm{T}}$. The matrices Ω_j, $j = 1, 2, \cdots$, are calculated forward as

$$\Omega_j = \Phi_{2^n}^{\mathrm{T}}(I_{2^n} \otimes \tilde{H}^{\mathrm{T}})\tilde{L}(\Theta_{j-1} \otimes I_{2^{m+p}}) \tag{3.46}$$

where Θ_{j-1} is obtained from the matrix Ω_{j-1} by deleting those columns with at most one non-zero entry. A non-negative integer r can be found so that there is at most one non-zero entry in each column of the matrix Ω_r.

(b) The matrix $\Phi_{2^n}^{\mathrm{T}}(I_{2^n} \otimes \tilde{H}^{\mathrm{T}})\tilde{L}$ contains only columns with at most one non-zero entry.

Proof. At first, it is shown that if the condition (a) is satisfied, then with knowledge of any admissible input and output trajectory of length $r+1$, the state estimate will converge to the correct state at time $0 \le t \le r$. Initializing Ω_0 as $\tilde{H}^{\mathrm{T}}$, one gets that $Col_j(\Omega_0)$, $j = 1, 2, \cdots, 2^p$ contains information of all possible states that may generate the output $\delta_{2^p}^j$. If $Col_j(\Omega_0) = \mathbf{0}_{2^n}$, then the output $y(0) = \delta_{2^p}^j$ is not admissible. If $Col_j(\Omega_0) \in \Delta_{2^n}$, then the state at time 0 can be uniquely determined. If there is more than one non-zero entry in $Col_j(\Omega_0)$, then the output $y(0) = \delta_{2^p}^j$ can not uniquely determine the state at time 0. Hence, the columns in the matrix Ω_0 belonging to $\Delta_{2^n} \bigcup \{\mathbf{0}_{2^n}\}$ are deleted and the result is denoted as Θ_0. Denote $\hat{x}(t)$ as state estimate at time t. $u(t)$ and $y(t+1)$ are the input and the output. The state estimate $\hat{x}(t+1)$ that may generate the output $y(t+1)$ can be determined by applying (3.41) and the element-wise multiplication (2.11). According to Lemma 2.10, one has

$$\begin{aligned}\hat{x}(t+1) &= \left(\tilde{L}\hat{x}(t)u(t)\right) \odot \left(\tilde{H}^{\mathrm{T}}y(t+1)\right)\\ &= \Phi_{2^n}^{\mathrm{T}}(I_{2^n} \otimes \tilde{H}^{\mathrm{T}})\tilde{L}\hat{x}(t)u(t)y(t+1).\end{aligned} \tag{3.47}$$

As $\hat{x}(t), u(t), y(t+1)$ are all column vectors, according to (2.1) there is $\hat{x}(t)u(t)y(t+1) = \hat{x}(t) \otimes u(t) \otimes y(t+1)$. Thus, each column of the matrix $\Theta_0 \otimes I_{2^{m+p}}$ corresponds to one combination of the not uniquely determined estimate $\hat{x}(0)$, the input $u(0) \in \Delta_{2^m}$ and the output $y(1) \in \Delta_{2^p}$. By comparing (3.47) with (3.46), it can be seen that the matrix Ω_1 contains the new state estimation at step 1. If some columns of the matrix Ω_1 are equal to $\mathbf{0}_{2^n}$, then the corresponding input and output trajectory is not admissible. If some columns of the matrix Ω_1 belong to Δ_{2^n}, then the state estimate already converges to the correct state. Hence, these columns are deleted. Repeating the same procedure, if a non-negative integer r can be found, such that all columns of the matrix Ω_r can be deleted, then the state estimate will converge to the correct state within r steps.

However, the condition (a) can not guarantee that after convergence the state estimate is still equal to the correct state. Therefore, the condition (b) will also be satisfied. As state estimate converges to the correct state, there is a positive index τ so that $\hat{x}(t) = x(t), \forall t \leq \tau$. According to (3.47), if the matrix $\Phi_{2^n}^{\mathrm{T}}(I_{2^n} \otimes \tilde{H}^{\mathrm{T}})\tilde{L}$ contains only columns with at most one non-zero entry, then for a given admissible input $u(t)$ and output $y(t+1)$ the vector $\Phi_{2^n}^{\mathrm{T}}(I_{2^n} \otimes \tilde{H}^{\mathrm{T}})\tilde{L}x(t)u(t)y(t+1)$ belongs to Δ_{2^n} which shows that the state estimate $\hat{x}(t+1)$ is equal to $x(t+1)$. □

Note that in a similar way as Theorem 3.15, the minimal index r that satisfies the conditions a–b in Theorem 3.19 is called the minimal decouplability index, denoted by r'_{min}. This index r'_{min} tells that after r'_{min} steps the knowledge of any admissible input and output trajectory can uniquely determine the current state.

3.3.3 Example

In order to show how to check the unknown input decouplability, consider the following BCN, which is a Boolean model for oxidative stress response pathways proposed in Sridharan et al. (2012),

$$\begin{cases} X_1(t+1) = U(t) \wedge \neg X_6(t), \\ X_2(t+1) = \neg X_1(t), \\ X_3(t+1) = \neg X_1(t) \wedge (X_5(t) \vee X_3(t)), \\ X_4(t+1) = X_2(t) \wedge \neg X_6(t), \\ X_5(t+1) = X_4(t) \vee \neg X_3(t), \\ X_6(t+1) = X_5(t) \wedge (\neg X_6(t) \vee \neg X_2(t)), \end{cases} \tag{3.48}$$

where X_1 represents the abundance of a biochemical entity called reactive oxidative species. X_2, X_3, X_4 and X_5 denote the abundance of certain proteins. X_6 represents the activation of the antioxidant genes. U is the oxidative stress signal. Assume that the abundance of the proteins X_2, X_3, X_5 can be directly measured, i.e.

$$\begin{cases} Y_1(t) = X_2(t), \\ Y_2(t) = X_3(t), \\ Y_3(t) = X_5(t). \end{cases} \tag{3.49}$$

Assume that the state X_5 is influenced by an unknown input $\xi(t)$ as $X_5(t+1) = (X_4(t) \vee \neg X_3(t)) \wedge \xi(t)$. An artificial unknown input $\omega(t)$ is introduced into the output equation $Y_1(t) = X_2(t)$ which becomes $Y_1(t) = X_2(t) \vee \omega(t)$.

To check the unknown input decouplability, let $\tilde{L} = L\mathbf{1}_2$ and $\tilde{H} = H\mathbf{1}_2$. The matrix $\Phi_n^{\mathrm{T}}(I_{2^n} \otimes \tilde{H}^{\mathrm{T}})\tilde{L}$ contains only the columns with at most one non-zero entry. Furthermore, recursively applying (3.46), it can be obtained that all columns of the matrix Ω_5 contain at most one non-zero entry. Hence, according to Theorem 3.19 the unknown inputs ξ and ω are decouplable. The minimal decouplability index is 5.

3.4 Reconstructibility of Large-scale Boolean Control Networks

Although the reconstructibility problem of BCNs can be solved, the main issue is still the computational complexity. Zhang et al. (2016) pointed out that the reconstructibility problem of BNCs is $\mathcal{NP}$-hard in general. In order to solve the reconstructibility problem of BCNs, necessary and sufficient conditions are known, for instance the result introduced in Section 3.2 and the conditions given in Fornasini and Valcher (2013), Xu and Hong (2013), Zhang et al. (2016). However, it is difficult to apply the existing approaches to large-scale BCNs because their computational complexities depend on the factor 2^n, where n is the number of state variables in the BCNs.

In the following parts, the reconstructibility problem of large-scale BCNs will be studied. For this purpose, a BCN is first partitioned into several subnetworks. Then, the reconstructibility problem of large-scale BCNs with a special graph structure, i.e. an acyclic structure, will be solved. After that, based on the unknown input decouplability, a sufficient condition for the reconstructibility analysis of general large-scale BCNs is given.

3.4.1 Subnetworks

For a BCN described by (2.34)–(2.35), the sets of states, inputs and outputs are denoted, respectively, by $\mathcal{X} = \{X_1, X_2, \cdots, X_n\}, \mathcal{U} = \{U_1, U_2, \cdots, U_m\}, \mathcal{Y} = \{Y_1, Y_2, \cdots, Y_p\}$. The nodes of each set can be partitioned into $\alpha \in \mathbb{Z}_+$ sets. That is

$$
\begin{aligned}
\mathcal{X} &= \mathcal{X}_1 \cup \mathcal{X}_2 \cup \cdots \cup \mathcal{X}_\alpha, \\
\mathcal{U} &= \mathcal{U}_1 \cup \mathcal{U}_2 \cup \cdots \cup \mathcal{U}_\alpha, \\
\mathcal{Y} &= \mathcal{Y}_1 \cup \mathcal{Y}_2 \cup \cdots \cup \mathcal{Y}_\alpha,
\end{aligned} \tag{3.50}
$$

where $\mathcal{X}_i = \{X_{i1}, X_{i2}, \cdots, X_{in_i}\}$, $\mathcal{U}_i = \{U_{i1}, U_{i2}, \cdots, U_{im_i}\}$ and $\mathcal{Y}_i = \{Y_{i1}, Y_{i2}, \cdots, Y_{ip_i}\}$ are, respectively, the subset of $\mathcal{X}, \mathcal{U}$ and $\mathcal{Y}$. $\mathcal{X}_i \cap \mathcal{X}_j$ and $\mathcal{Y}_i \cap \mathcal{Y}_j$ are empty for $i \neq j$, $\sum_{i=1}^{\alpha} n_i = n$ and $\sum_{i=1}^{\alpha} p_i = p$.

In graph theory, a directed graph can be denoted by $G = (\mathcal{V}, \mathcal{E})$ where $\mathcal{V}$ is a set of nodes and $\mathcal{E}$ is a set of edges. Suppose that $(i, j) \in \mathcal{E}$ holds for some nodes $i, j \in \mathcal{V}$. Then, the vertex i is said to be an in-neighbor of j (Zhu and Jiang, 2015). A block defined as $\mathcal{N}_i = \mathcal{X}_i \cup \mathcal{U}_i \cup \mathcal{Y}_i$ is called a super node. The in-neighbors of subnetwork $\mathcal{N}_i$ are defined as follows.

Definition 3.20 (In-Neighbors) A block $\mathcal{N}_i$ has incoming edges from some states in $\mathcal{X}_j, j \neq i$. Then, the block $\mathcal{N}_j, j \neq i$ is called an in-neighbor of the block $\mathcal{N}_i$.

Let the i-th subsystem consist of the nodes in block $\mathcal{N}_i$. According to Definition 3.20, some states of in-neighbors can be interpreted as the input for the i-th subnetwork. The set of these state variables of in-neighbors for each block $\mathcal{N}_i$ is denoted by $\mathcal{Z}_i = \{Z_{iq_1}, Z_{iq_2}, \cdots, Z_{iq_i}\} \subseteq \mathcal{X} \backslash \mathcal{X}_i$.

Suppose that large-scale BCN is partitioned into subnetworks and the output variables $Y_{ij},\ j = 1, 2, \cdots, p_i$ are local measurements of the i-th subsystem. The dynamic of the i-th subsystem containing the nodes of the i-th block $\mathcal{N}_i$, $\forall i = 1, 2, \cdots, \alpha$ can be described as

$$
\begin{cases}
X_{i1}(t+1) = f_{i1}(X_{i1}(t), X_{i2}(t), \cdots, X_{in_i}(t), U_{i1}(t), \cdots, U_{im_1}(t), Z_{i1}(t), \cdots, Z_{iq_i}(t)), \\
\quad \vdots \\
X_{in_i}(t+1) = f_{in_i}(X_{i1}(t), X_{i2}(t), \cdots, X_{in_i}(t), U_{i1}(t), \cdots, U_{im_i}(t), Z_{i1}(t), \cdots, Z_{iq_i}(t)), \\
Y_{i1}(t) = h_{i1}(X_{i1}(t), X_{i2}(t), \cdots, X_{in_i}(t)), \\
\quad \vdots \\
Y_{ip_i}(t) = h_{ip_i}(X_{i1}(t), X_{i2}(t), \cdots, X_{in_i}(t)).
\end{cases} \tag{3.51}
$$

In a similar way as Section 2.3, the i-th subnetwork (3.51) can be expressed in a matrix form as

$$x_{sub_i}(t+1) = L_{sub_i} x_{sub_i}(t) u_{sub_i}(t) z_i(t), \tag{3.52}$$

$$y_{sub_i}(t) = H_{sub_i} x_{sub_i}(t), \tag{3.53}$$

where $x_{sub_i}(t) = \ltimes_{j=1}^{n_i} x_{ij}(t) \in \Delta_{2^{n_i}}$, $u_{sub_i}(t) = \ltimes_{j=1}^{m_i} u_{ij}(t) \in \Delta_{2^{m_i}}$, $y_{sub_i}(t) = \ltimes_{j=1}^{p_i} y_{ij}(t) \in \Delta_{2^{p_i}}$, $z_i(t) = \ltimes_{j=1}^{q_i} z_{ij}(t) \in \Delta_{2^{q_i}}$ are, respectively, the state, the input, the output and the influence of the in-neighbors. $L_{sub_i} \in \mathcal{L}_{2^{n_i} \times 2^{n_i+m_i+q_i}}$, $H_{sub_i} \in \mathcal{L}_{2^{p_i} \times 2^{n_i}}$ are the logical matrices that contain all the structural information of the logical functions in (3.51).

Note that if no in-neighbor exists, then $\mathcal{Z}_i = \emptyset$. In this case, (3.52) is changed to $x_{sub_i}(t+1) = L_{sub_i} x_{sub_i}(t) u_{sub_i}(t)$. Furthermore, due to $\mathcal{X}_i \cap \mathcal{X}_j = \emptyset$ and $\mathcal{Y}_i \cap \mathcal{Y}_j = \emptyset$ for $i \neq j$, one gets $x(t) = x_{sub_1}(t) x_{sub_2}(t) \cdots x_{sub_\alpha}(t)$ and $y(t) = y_{sub_1}(t) y_{sub_2}(t) \cdots y_{sub_\alpha}(t)$. It is worth mentioning that the partitioning of a large-scale BCN is not unique (Zhao et al., 2016). In order to achieve a good partition, the following principles should be considered:

1. Each subnetwork should contain a small number of nodes to reduce the computational effort.
2. Each subnetwork should connect to a small number of in-neighbors subsystems to reduce communication.

Remark 3.21 Different from the aggregation of BCNs described in Zhao et al. (2016), output variables are considered. Furthermore, subsystems are allowed to be influenced by the same input variables.

3.4.2 Reconstructibility of Large-scale Boolean Control Networks With Acyclic Structure

In mathematics and computer science, a directed acyclic graph is composed of variables (nodes) and arrows between nodes (directed edges) so that no directed cycles exist in the graph (Pahl and Damrath, 2001).

Assume that a large-scale BCN described by (2.34)–(2.35) is partitioned into α subnetworks described by (3.52)–(3.53). Every directed acyclic graph has a topological ordering. Hence, all the subnetworks of the large-scale BCN can be sorted into β levels denoted by Ξ_j, $j = 0, 1, \cdots, \beta - 1$. Consider each subnetwork as a super node. If the large-scale BCN shows a structure of a directed acyclic graph (i.e. acyclic structure), then the following result can be obtained.

Theorem 3.22 *A large-scale BCN described by (2.34)–(2.35) with acyclic structure is reconstructible if all subnetworks are reconstructible.*

Proof. Subnetworks are sorted into β levels. The subnetworks at level i have no income-edge from the subnetworks at level $j \geq i$ with $i, j \in \{0, 1, \cdots, \beta - 1\}$. Note that subnetworks at level 0 do not have any in-neighbor. As all subnetworks at level 0 are reconstructible according to Definition 3.5, an integer t_0 can be found so that the states of all subnetworks at level 0 are correctly determined at time t_0. For the subnetworks at level 1, they can be influenced by the states of their in-neighbors at level 0. Once the states of the in-neighbors at level 0 are correctly estimated, the inputs for all subnetworks at level 1 are also known. Therefore, an integer $t_1 \geq t_0$ can be found so that the states of all subnetworks at level 1 are correctly estimated at time t_1. Repeating the procedure until the last level, the result is obtained. □

Remark 3.23 Assume that the minimal reconstructibility index of the i-th subnetwork is $r_{min,i}$, $i = 1, 2, \cdots, \alpha$. Then the integer t_j can be calculated by

$$t_0 = \max_{i \in \Xi_0} r_{min,i} \quad \text{and} \quad t_j = t_{j-1} + \max_{i \in \Xi_j} r_{min,i}, \tag{3.54}$$

so that any admissible input and output trajectory $\{(y(t), u(t)), t = 0, 1, \cdots, t_j\}$ can correctly determine the state $x(t_j)$ of subnetworks at level j. This means that the minimal reconstructibility index r_{min} of the large-scale BCN is at most $t_{\beta-1}$.

3.4.3 Reconstructibility of General Large-scale Boolean Control Networks

In this subsection, the reconstructibility problem of general large-scale BCNs shall be studied. Let the large-scale BCNs be partitioned into several subnetworks. These subnetworks usually build a directed graph with more general structure (i.e. both cyclic and acyclic structure). The aim is to introduce a sufficient condition, under which general large-scale BCNs with cyclic structure are reconstructible.

Recall that a subnetwork can be described by (3.52)–(3.53) where the state $z_i(t)$ is added to describe the influence of the in-neighbors. However, if the current state of a subnetwork can be uniquely determined by the knowledge of input and output trajectory no matter what the state $z_i(t)$ is, then the influence of in-neighbors can be ignored. Therefore, the edges between the subnetwork and the in-neighbors in the graph can be cut. By cutting the edges, the large-scale BCN has an acyclic structure.

Hence, the approach proposed in Section 3.4.2 can be applied to the reconstructibility analysis of large-scale BCNs with cyclic structure.

Applying Proposition 2.4, a subnetwork (3.52)–(3.53) can be equivalently written as

$$x_{sub_i}(t+1) = L_{sub_i} W_{[2^{q_i}, 2^{n_i+m_i}]} z_i(t) x_{sub_i}(t) u_{sub_i}(t), \tag{3.55}$$

$$y_{sub_i}(t) = H_{sub_i} x_{sub_i}(t). \tag{3.56}$$

If the state $z_i(t)$ is considered as the unknown input $\xi(t)$ and the unknown input $\omega(t)$ is set to 1 (i.e. no unknown input in the output equation), then the subnetwork (3.55)–(3.56) is actually a BCN with unknown inputs (3.31)–(3.32). Therefore, in order to check the unknown input decouplability of the subnetwork described by (3.55)–(3.56), a new necessary and sufficient condition which is a slightly modified version of Theorem 3.19.

Theorem 3.24 *Let* $\tilde{L}_{sub_i} = L_{sub_i} W_{[2^{q_i}, 2^{n_i+m_i}]} \mathbf{1}_{2^{q_i}}$. *A BCN* (3.55)–(3.56) *is unknown input decouplable with respect to the unknown input* $z_i(t)$ *if and only if*

(a) initialize $\Omega_0 = H^{\mathrm{T}}_{sub_i}$. *The matrices* Ω_j, $j = 1, 2, \cdots$, *are calculated forward as*

$$\Omega_j = \Phi^{\mathrm{T}}_{2^{n_i}} (I_{2^{n_i}} \otimes H^{\mathrm{T}}_{sub_i}) \tilde{L}_{sub_i} (\Theta_{j-1} \otimes I_{2^{m_i+p_i}}) \tag{3.57}$$

where Θ_{j-1} *is obtained from the matrix* Ω_{j-1} *by deleting those columns with at most one non-zero entry. A non-negative integer* r *can be found so that there is at most one non-zero entry in each column of the matrix* Ω_r.

(b) the matrix $\Phi^{\mathrm{T}}_{2^{n_i}} (I_{2^{n_i}} \otimes H^{\mathrm{T}}_{sub_i}) \tilde{L}_{sub_i}$ *contains only columns with at most one non-zero entry.*

Assume that a subnetwork in the large-scale BCN is unknown inputs decouplable, then a sufficient condition for reconstructibility of general large-scale BCNs can be given as follows.

Theorem 3.25 *A large-scale BCN with cyclic structure is reconstructible if all of the following conditions are satisfied:*

1. *All the subnetworks are reconstructible.*
2. *There is a set* Ψ *of subnetworks* (3.52)–(3.53). *The* i*-th subnetwork is unknown input decouplable with respect to the input* z_i, $\forall i \in \Psi$.

3. *Cutting the income-edges of all subnetworks contained in the set Ψ, the resulting graph belongs to directed acyclic graph.*

Proof. According to Definition 3.18, the current state of unknown input decouplable subnetworks can be uniquely determined by the input and output trajectory without information coming through income-edge. Therefore, for state reconstruction these income-edges can be cut off. If the resulting graph belongs to a directed acyclic graph and all subnetworks are reconstructible, then the large-scale BCN is reconstructible according to Theorem 3.22. □

Note that the same result to estimate the upper bound of minimal reconstructibility index of general large-scale BCN as Remark 3.23 can be obtained.

3.4.4 Example

In order to illustrate the main results in this section, a slight modification of the model in Zhao et al. (2016) will be taken into consideration. Assume that the states X_3, X_4 and X_8 can be directly measured. The dynamics of the modified BCN can be described by

$$\begin{cases} X_1(t+1) = X_2(t), \\ X_2(t+1) = X_3(t) \wedge X_7(t), \\ X_3(t+1) = X_1(t) \wedge X_2(t) \wedge U_1(t), \\ X_4(t+1) = \neg X_1(t) \wedge X_5(t) \wedge X_7(t), \\ X_5(t+1) = X_4(t) \wedge U_2(t), \\ X_6(t+1) = X_6(t) \vee X_8(t), \\ X_7(t+1) = X_6(t), \\ X_8(t+1) = X_5(t) \vee X_7(t) \vee X_9(t), \\ X_9(t+1) = \neg X_8(t), \\ Y_1(t) = X_3(t), \quad Y_2(t) = X_4(t), \quad Y_3(t) = X_8(t). \end{cases} \tag{3.58}$$

The sets of states, inputs and outputs are, respectively,

$$\begin{cases} \mathcal{X} = \{X_1, X_2, X_3, X_4, X_5, X_6, X_7, X_8, X_9\}, \\ \mathcal{U} = \{U_1, U_2\}, \\ \mathcal{Y} = \{Y_1, Y_2, Y_3\}, \end{cases} \tag{3.59}$$

and all elements in $\mathcal{X} \cup \mathcal{U} \cup \mathcal{Y}$ are nodes of the directed graph shown in Fig. 3.2.

Subnetworks

Now the sets $\mathcal{X}, \mathcal{U}$ and $\mathcal{Y}$ are, respectively, partitioned into three subsets. The states, inputs and outputs of each subnetwork are (see Fig. 3.2)

$$\begin{aligned} &\mathcal{X}_1 = \{X_1, X_2, X_3\}, && \mathcal{U}_1 = \{U_1\}, && \mathcal{Y}_1 = \{Y_1\}, \\ &\mathcal{X}_2 = \{X_4, X_5\}, && \mathcal{U}_2 = \{U_2\}, && \mathcal{Y}_2 = \{Y_2\}, \\ &\mathcal{X}_3 = \{X_6, X_7, X_8, X_9\}, && \mathcal{U}_3 = \emptyset, && \mathcal{Y}_3 = \{Y_3\}. \end{aligned} \tag{3.60}$$

Hence, the dynamics of three subnetworks are described by

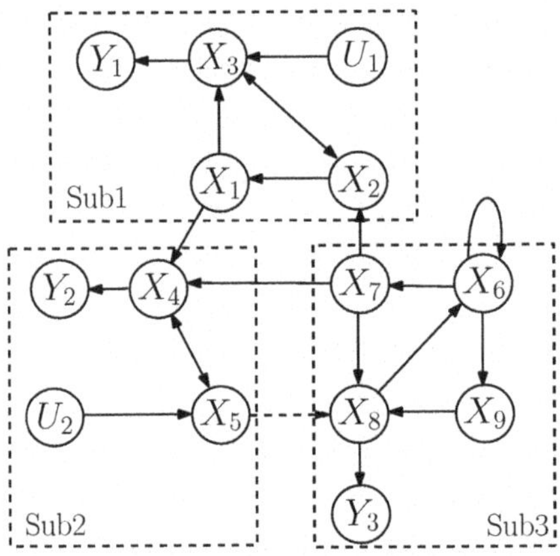

Figure 3.2 An example of BCN consisting of three subnetworks

$$Sub1 : \begin{cases} X_1(t+1) &= X_2(t), \\ X_2(t+1) &= X_3(t) \wedge X_7(t), \\ X_3(t+1) &= X_1(t) \wedge X_2(t) \wedge U_1(t), \\ Y_1(t) &= X_3(t); \end{cases} \tag{3.61}$$

$$Sub2 : \begin{cases} X_4(t+1) &= \neg X_1(t) \wedge X_5(t) \wedge X_7(t), \\ X_5(t+1) &= X_4(t) \wedge U_2(t), \\ Y_2(t) &= X_4(t); \end{cases} \tag{3.62}$$

$$Sub3 : \begin{cases} X_6(t+1) &= X_6(t) \vee X_8(t), \\ X_7(t+1) &= X_6(t), \\ X_8(t+1) &= X_5(t) \vee X_7(t) \vee X_9(t), \\ X_9(t+1) &= \neg X_8(t), \\ Y_3(t) &= X_8(t). \end{cases} \tag{3.63}$$

According to Section 3.4.1, the subsystems $Sub1$, $Sub2$ and $Sub3$ can be described in the following algebraic form:

$$\begin{aligned} x_{sub_i}(t+1) &= L_{sub_i} x_{sub_i}(t) u_{sub_i}(t) z_i(t), \\ y_{sub_i}(t) &= H_{sub_i} x_{sub_i}(t), \quad i = 1, 2, 3. \end{aligned} \tag{3.64}$$

From Fig. 3.2, it can be recognized that the subnetwork $Sub1$ has only one in-neighbor, i.e. the subnetwork $Sub3$. Therefore, let $x_{sub_1}(t) = x_1(t)x_2(t)x_3(t)$, $u_{sub_1}(t) = u_1(t)$, $y_{sub_1}(t) = y_1(t)$, $z_1(t) = x_7(t)$. The logical matrices $L_{sub_1} \in \mathcal{L}_{8\times 32}$, $H_{sub_1} \in \mathcal{L}_{2\times 8}$ are

$$L_{sub_1} = \delta_8[1\ 3\ 2\ 4\ 3\ 3\ 4\ 4\ 6\ 8\ 6\ 8\ 8\ 8\ 8\ 8\ 2\ 4\ 2\ 4\ 4\ 4\ 4\ 4\ 6\ 8\ 6\ 8\ 8\ 8\ 8\ 8],$$

$$H_{sub_1} = \delta_2[1\ 2\ 1\ 2\ 1\ 2\ 1\ 2].$$

In the same way, the subnetworks $Sub1$ and $Sub3$ are the in-neighbors of the subnetwork $Sub2$. So, there are $x_{sub_2}(t) = x_4(t)x_5(t)$, $u_{sub_2}(t) = u_2(t)$, $y_{sub_2}(t) = y_2(t)$, $z_2(t) = x_1(t)x_7(t)$. The logical matrices $L_{sub_2} \in \mathcal{L}_{4\times 32}$, $H_{sub_2} \in \mathcal{L}_{2\times 4}$ are

$$L_{sub_2} = \delta_4[3\ 3\ 1\ 3\ 4\ 4\ 2\ 4\ 3\ 3\ 3\ 3\ 4\ 4\ 4\ 4\ 4\ 4\ 2\ 4\ 4\ 4\ 2\ 4\ 4\ 4\ 4\ 4\ 4\ 4\ 4\ 4],$$

$$H_{sub_2} = \delta_2[1\ 1\ 2\ 2].$$

For the subnetwork $Sub3$, the state variable X_5 can influence subnetwork $Sub3$. Hence, $x_{sub_3}(t) = x_6(t)x_7(t)x_8(t)x_9(t)$, $z_3 = x_5(t)$. The logical matrices $L_{sub_3} \in \mathcal{L}_{16\times 32}$, $H_{sub_3} \in \mathcal{L}_{2\times 16}$ are

$$L_{sub_3} = \delta_{16}[2\ 2\ 2\ 2\ 1\ 1\ 1\ 1\ 2\ 2\ 2\ 4\ 1\ 1\ 1\ 3\ 6\ 6\ 6\ 6\ 13\ 13\ 13\ 13\ 6\ 6\ 6\ 8\ 13\ 13\ 13\ 13],$$

$$H_{sub_3} = \delta_2[1\ 1\ 2\ 2\ 1\ 1\ 2\ 2\ 1\ 1\ 2\ 2\ 1\ 1\ 2\ 2].$$

Reconstructibility analysis

Next, the reconstructibility of the BCN (3.58) will be checked. Considering the three subnetworks (3.61)–(3.63) as super nodes in Fig. 3.2, the BCN (3.58) shows a cyclic structure, i.e. a directed cyclic graph. Therefore, Theorem 3.25 can be applied.

At first, reconstructibility of the subnetwork $Sub1$ will be analyzed. The matrix $\Omega_{1,0}$ is initialized as $H^{\mathrm{T}}_{sub_1}$. As no column should be deleted, the matrix $\Omega_{1,1}$ is calculated according to (3.22). It is found out that all the columns of $\Omega_{1,1}$ are one of the column vectors $\mathbf{0}_8$, δ_8^1, δ_8^3, $\delta_8^4 + \delta_8^8$, $\delta_8^2 + \delta_8^6$. According to Theorem 3.15, the columns in the matrix $\Omega_{1,1}$ equal to $\mathbf{0}_8$, δ_8^1 and δ_8^3 should be deleted. Then, again based on (3.22), the matrix $\Omega_{1,2}$ is calculated. Because all columns of the matrix $\Omega_{1,2}$ contain at most one non-zero entry, the subnetwork $Sub1$ is reconstructible and its minimal reconstructibility index is 2.

In the next step, reconstructibility of the subnetworks $Sub2$, $Sub3$ is checked in the same way as described above. The subnetworks $Sub2$, $Sub3$ are reconstructible and their minimal reconstructibility indices are, respectively, 1 and 4.

Additionally, unknown input decouplability of the subnetwork $Sub3$ is checked. Let $\tilde{L}_{sub_3} = L_{sub_3} W_{[2,16]} \mathbf{1}_2$. A simple calculation shows that the matrix $\Phi^{\mathrm{T}}_{16}(I_{16} \otimes H^{\mathrm{T}}_{sub_3})\tilde{L}_{sub_3}$ has only columns with at most one non-zero entry. Next, the matrix $\tilde{\Omega}_{3,0}$ is initialized as $H^{\mathrm{T}}_{sub_3}$. Applying (3.22) recursively, it is found out that all columns of the matrix $\tilde{\Omega}_{3,4}$ contain at most one non-zero entry. According to Theorem 3.24, the subnetwork $Sub3$ is unknown input reconstructible and the income-edge $X_5 \rightarrow X_8$ in Fig. 3.2 can be cut off. After that, the modified directed graph in Fig. 3.2 shows an acyclic structure with $Sub3$ at level 0, $Sub1$ at level 1 and $Sub2$ at level 2.

As a result, the BCN (3.58) is reconstructible according to Theorem 3.25. Based on the results obtained above, it can also be concluded that the minimal reconstructibility index of the BCN (3.58) has an upper bound of 7.

Observer Design

4

In this chapter, observer design of BCNs will be considered. At first, the Luenberger-like observer design approach for BCNs is proposed in Section 4.1. Based on the Luenberger-like observer and by taking all possible unknown inputs into account with the help of the all-ones column vector, an approach to design unknown input observers for BCNs is proposed in Section 4.2. In addition, distributed observer design and reduced-order observer design will be considered, respectively, in Section 4.4 and 4.3 to facilitate an online implementation for large-scale BCNs.

BCNs considered in this chapter are described in the form (2.34)–(2.35). The Boolean product (2.36) will be applied. For the sake of simplicity, the symbol $\ltimes_{\mathcal{B}}$ will be omitted.

4.1 Luenberger-like Observers

In this section, an approach for Luenberger-like observer design will be proposed. In addition, the performance of the proposed observer will be studied. It will be shown that the real state is always contained in the state estimate provided by the proposed observer and for a reconstructible BCN the state estimate will converge to the real state.

Recall that the minimal reconstructibility index is denoted as r_{min}. Recently, Shift-Register observer for state estimation of BCNs has been proposed in Fornasini and Valcher (2013) based on the mapping that associates the knowledge of admissible input and output trajectories $\{y(t-r_{min}), u(t-r_{min}), y(t-r_{min}+1), u(t-r_{min}+1), \cdots, y(t), u(t)\}$ to the current state $x(t)$ according to reconstructibility in the sense of Definition 3.7 directly.

Z. Zhang, *Observer Design for Control and Fault Diagnosis of Boolean Networks*,
https://doi.org/10.1007/978-3-658-35929-4_4

Lemma 4.1 (Fornasini and Valcher (2013)). Initialize the initial state of the Shift-Register observer as $z(r_{min}) = y(0)u(0)y(1)u(1)\cdots y(r_{min})u(r_{min})$. The Shift-Register observer for the BCN (2.34)–(2.35) is

$$\hat{x}(t) = \hat{H}z(t), \tag{4.1}$$
$$z(t+1) = (\mathbf{1}^{\mathrm{T}}_{2^{m+p}} \otimes I_{2^{m+p}})W_{[2^{m+p},2^{(r+1)(m+p)}]}y(t+1)u(t+1)z(t),\ t = r_{min}, r_{min}+1, \cdots$$

where $\hat{H} \in \mathcal{L}_{2^n \times 2^{(r_{min}+1)(m+p)}}$ is the system matrix, $\hat{x}$ is estimate of the BCN state, $y(t+1), u(t+1)$ are, respectively the output and the input at time $t+1$. Beginning at time $t = 0$, the Shift-Register observer can always provide the correct state estimate, i.e. $\hat{x}(t) = x(t)$, at time $t \geq r_{min}$.

Note that $z(t)$ is the vector expression of input and output trajectory $y(t - r_{min})u(y - r_{min})y(t - r_{min} + 1)u(y - r_{min} + 1)\cdots y(t)u(t)$ and $\hat{H}$ associates every admissible input and output trajectory described by $z(t)$ with the unique state $x(t)$ called reconstructibility map (Fornasini and Valcher, 2013).

However, in Fornasini and Valcher (2013), the matrix $\hat{H}$ is only determined according to the reconstructibility map that associates every admissible input and output trajectory of the length $r_{min} + 1$ with the unique state $x(t)$. No mathematical formula is given to directly calculate the matrix $\hat{H}$. Therefore, in the following, a mathematical formula will be given to calculate the matrix $\hat{H}$.

By looking at (3.5) and (3.12), it can be recognized that replacing r with r_{min}, the matrix $\hat{H}$ in (4.1) has the same structure as $\Omega_{r_{min}}$. Due to the Boolean product as defined in (2.36), each column of the matrix Ω_r belongs to the set $\Delta_{2^n} \cup \{\mathbf{0}_{2^n}\}$. Let Ω_r^* be a logical matrix converted from Ω_r by replacing the zero columns with $\delta^1_{2^n}$. If the matrix $\hat{H}$ is set to

$$\hat{H} = \Omega_{r_{min}} \otimes \mathbf{1}^{\mathrm{T}}_{2^m}, \tag{4.2}$$

then the following result can be achieved.

Theorem 4.2 *If the BCN (2.34)–(2.35) is reconstructible and the Shift-Register observer* (4.1) *is used with* $\hat{H}$ *given in* (4.2) *to generate the state estimate* $\hat{x}(t)$ *for* $t \geq r_{min}$, *then the state of the BCN is provided by the observer without estimation error, i.e.,* $\hat{x}(t) = x(t),\ \forall t \geq r_{min}$.

Proof. Recall that $u(r_{min})$ is only used in the transition from $x(r_{min})$ to $x(r_{min} + 1)$ and $1 = \mathbf{1}^{\mathrm{T}}_{2^m}u(r_{min})$ always holds. Therefore, the right side of (3.12) can be multiplied with $\mathbf{1}_{2^m}^{\mathrm{T}} u(r_{min})$ which does not change the equality. So (3.12) can be equivalently rewritten as

$$\begin{aligned}\hat{x}(r_{min}) &= \Omega_{r_{min}} \left(\prod_{j=0}^{r_{min}-1} y(j)u(j) \right) y(r_{min}) \mathbf{1}_{2^m}^{\mathrm{T}} u(r_{min}) \\ &= \Omega_{r_{min}} (I_{2^{r_{min}\cdot(m+p)+p}} \otimes \mathbf{1}_{2^m}^{\mathrm{T}}) \left(\prod_{j=0}^{r_{min}} y(j)u(j) \right) \\ &= \left(\Omega_{r_{min}} \otimes \mathbf{1}_{2^m}^{\mathrm{T}}\right) \left(\prod_{j=0}^{r_{min}} y(j)u(j) \right).\end{aligned} \tag{4.3}$$

The state $x(r_{min})$ that is compatible with the input and output trajectory $\{y(0), u(0), y(1), u(1), \cdots, y(r_{min}), u(r_{min})\}$ can be determined by (4.3). Hence, the matrix $\Omega_{r_{min}} \otimes \mathbf{1}_{2^m}^{\mathrm{T}}$ corresponds to the reconstructibility map. As $\hat{H}$ determined by (4.2), the reconstructibility map still holds. After replacing the zero columns of the matrix $\hat{H}$ by $\delta_{2^n}^1$, the modified matrix $\hat{H}$ is equal to the static map shown in Fornasini and Valcher (2013). According to Lemma 4.1, the Shift-Register observer with $\hat{H}$ determined by (4.2) produces an estimate equal to the correct state, i.e. $\hat{x}(t) = x(t)$ for $t \geq r_{min}$. □

Note that in the Shift-Register observer (4.1) high computational effort is required to multiply the high-dimensional matrix $\hat{H} \in \mathcal{L}_{2^n \times 2^{(r_{min}+1)(m+p)}}$ by vector $z(t) \in \mathcal{L}_{2^{(r_{min}+1)(m+p)} \times 1}$ at each time instance. This problem can be solved by introducing a new observer which is introduced as follows.

Theorem 4.3 *The Luenberger-like observer with initial estimated state $\hat{x}_s(0) = H^{\mathrm{T}} y(0)$ described by*

$$\begin{aligned}\hat{x}_s(t) &= \Phi_{2^n}^{\mathrm{T}} L_{eq} \hat{x}_s(t-1) u(t-1) H^{\mathrm{T}} y(t) \\ &= \Phi_{2^n}^{\mathrm{T}} (I_{2^n} \otimes H^{\mathrm{T}}) L_{eq} \hat{x}_s(t-1) u(t-1) y(t)\end{aligned} \tag{4.4}$$

provides the same state estimate as the Shift-Register observer (4.1) *at time $t \geq r_{min}$.*

Proof. At the beginning, the state estimate of the Luenberger-like observer at time $t = 0$ is initialized as $\hat{x}_s(0) = H^{\mathrm{T}} y(0)$. By recursive applying (4.4), one obtains

$$
\begin{aligned}
\hat{x}_s(r_{min}) &= \Phi_{2^n}^{\mathrm{T}} L_{eq} \hat{x}_s(r_{min}-1) u(r_{min}-1) H^{\mathrm{T}} y(r_{min}) \\
&= (\Phi_{2^n}^{\mathrm{T}} L_{eq})^2 \hat{x}_s(r_{min}-2) u(r_{min}-2) H^{\mathrm{T}} y(r_{min}-1) u(r_{min}-1) H^{\mathrm{T}} y(r_{min}) \\
&\ \ \vdots \\
&= (\Phi_{2^n}^{\mathrm{T}} L_{eq})^{r_{min}} \underbrace{\hat{x}_s(0)}_{=H^{\mathrm{T}} y(0)} u(0) \left(\prod_{j=1}^{r_{min}-1} H^{\mathrm{T}} y(j) u(j) \right) H^{\mathrm{T}} y(r_{min}) \qquad (4.5) \\
&= (\Phi_{2^n}^{\mathrm{T}} L_{eq})^{r_{min}} \left(\prod_{j=0}^{r_{min}-1} H^{\mathrm{T}} y(j) u(j) \right) H^{\mathrm{T}} y(r_{min}) \\
&= \left(\Phi_{2^n}^{\mathrm{T}} L_{eq} \right)^{r_{min}} \left(\prod_{i=0}^{r} (I_{2^{i \cdot (n+m)}} \otimes H) \right)^{\mathrm{T}} \left(\prod_{j=0}^{r_{min}-1} y(j) u(j) \right) y(r_{min}).
\end{aligned}
$$

Recall that the matrix $\Omega_{r_{min}}$ can be calculated by (3.18). Comparing (4.5) with (4.3), it is straightforward to get that $\hat{x}_s(r) = \hat{x}(r)$. Furthermore, due to the minimal reconstructibility index r_{min}, the Shift-Register observer with $\hat{H}$ determined by (4.2) provides state estimate at time $t \geq r_{min}$. Hence, the Shift-Register observer and Luenberger-like observer produce the same state estimate at time $t \geq r_{min}$. □

Unlike state observers for discrete-time linear systems, the initial state $\hat{x}(0)$ of the Luenberger-like observer (6.70) cannot be chosen arbitrarily, but shall be initialized as $\hat{x}(0) = H^{\mathrm{T}} y(0)$ with the output information $y(0)$. In this way, it can guarantee that the real state $x(t)$ $\forall t = 0, 1, \cdots$ is always contained in the set represented by $\hat{x}_s(t)$.

Theorem 4.4 *Initializing the state of the Luenberger-like observer* (4.4) *as* $\hat{x}_s(0) = H^{\mathrm{T}} y(0)$, *the set of states represented by* $\hat{x}_s(t), t = 0, 1, \cdots$ *always contains the real state* $x(t)$.

Proof. The theorem is proven by induction. At first, as the real state $x(0)$ generates the output $y(0) \in \Delta_{2^p}$, it is simple to get that the real state $x(0)$ is always contained in the set represented by $\hat{x}_s(0)$. Suppose that the set represented by $\hat{x}_s(t-1)$ contains the real state $x(t-1)$ at time $t-1$, it shall be shown that the real state $x(t)$ belongs to the set represented by $\hat{x}_s(t)$. Let $\bar{x}(t)$ be a vector representing the set of all possible states $x(t)$ that can generate the output $y(t)$. Recall that H is a logical matrix. The vector $\bar{x}(t)$ satisfies

$$\bar{x}(t) = H^{\mathrm{T}} y(t). \qquad (4.6)$$

As the real state $x(t)$ satisfies (2.35), it is clear that the real state $x(t)$ belongs to the set represented by $\bar{x}(t)$. According to Proposition 2.9, there is

$$x(t) = x(t) \odot \bar{x}(t) = \Phi_{2^n}^{\mathrm{T}} x(t)\bar{x}(t). \tag{4.7}$$

According to the assumption and the Luenberger-like observer (4.4), one has

$$\begin{aligned} \Phi_{2^n}^{\mathrm{T}}(I_{2^n} \otimes H^{\mathrm{T}})L_{eq}x(t-1)u(t-1)y(t) =& \Phi_{2^n}^{\mathrm{T}} L_{eq}x(t-1)u(t-1)H^{\mathrm{T}}y(t) \\ =& \Phi_{2^n}^{\mathrm{T}} L_{eq}x(t-1)u(t-1)\bar{x}(t) \\ =& \Phi_{2^n}^{\mathrm{T}} x(t)\bar{x}(t) \\ =& x(t) \odot \bar{x}(t) = x(t) \end{aligned} \tag{4.8}$$

which shows that if the real state $x(t-1)$ belongs to the set represented by $\hat{x}_s(t-1)$, then the set represented by $\hat{x}_s(t)$ contains the real state $x(t)$. □

Remark 4.5 The Luenberger-like observer (6.70) provides the correct state estimate (i.e. $\hat{x}(t) = x(t)$) not later than $t = r_{min}$.

As mentioned in Weiss and Margaliot (2019), state observers have exponential complexity. Now the complexity of the online computation of the Shift-Register observer and the Luenberger-like observer is analyzed. Recall that $\hat{H}$ is a matrix of dimension $2^n \times 2^{(r_{min}+1)\cdot(m+p)}$ and $z(t)$ is a vector of dimension $2^{(r_{min}+1)\cdot(m+p)} \times 1$. The multiplication $\hat{H}z(t)$ has the computational complexity of $\mathcal{O}(2^{n+(r_{min}+1)\cdot(m+p)})$. As the matrix $(\mathbf{1}_{2^{m+p}}^{\mathrm{T}} \otimes I_{2^{(r_{min}+1)\cdot(m+p)}})W_{[2^{m+p},2^{(r_{min}+1)\cdot(m+p)}]}$ has dimensions $2^{(r_{min}+1)\cdot(m+p)} \times 2^{(r_{min}+2)\cdot(m+p)}$, to update information of the input and output trajectory according to (4.1) the computational effort $\mathcal{O}(2^{(2r_{min}+3)\cdot(m+p)})$ is required. Hence, applying the Shift-Register observer for online state estimation, it runs in $\mathcal{O}(2^{(2r_{min}+3)\cdot(m+p)} + 2^{n+(r_{min}+1)\cdot(m+p)}))$ at each time instance. In contrast, $\hat{x}_s(t-1)u(t-1)y(t)$ is a vector of the dimension $2^{n+m+p} \times 1$. Applying the Luenberger-like observer for online state estimation, the $2^n \times 2^{n+m+p}$ dimensional matrix $\Phi_{2^n}^{\mathrm{T}}(I_{2^n} \otimes H^{\mathrm{T}})L_{eq}$ is multiplied with the vector $\hat{x}_s(t-1)u(t-1)y(t)$, which requires $\mathcal{O}(2^{2n+m+p})$ computational effort. As Fornasini and Valcher (2013) pointed out that the index r_{min} has an upper bound $\frac{(2^n+1)\cdot(2^n-2)}{2}$, the computational burden of the Shift-Register observer is $\mathcal{O}(2^{n+(\frac{(2^n+1)\cdot(2^n-2)}{2}+1)\cdot(m+p)} + 2^{((2^n+1)\cdot(2^n-2)+3)\cdot(m+p)})$, which is much higher than the Luenberger-like observer (i.e. $\mathcal{O}(2^{2n+m+p})$). Therefore, the Luenberger-like observer needs much less online calculation than the Shift-Register observer.

It is important to note that the Luenberger-like observer (4.4) estimates the state of BCNs in the same way as the observer given in Fornasini and Valcher (2015a) for fault detection (see Section IV.B). However, the observer in Fornasini and Valcher (2015a) is not described by means of a BCN, but by a two-step calculation procedure which makes it harder to evaluate its performance. Besides, the convergence of the observer proposed in Fornasini and Valcher (2015a) is not evaluated. For automata, a STP based approach for observer design was proposed in Xu and Hong (2013). But the relationship between the reconstructibility and the convergence of the observer is unknown.

State estimation by using the proposed Luenberger-like observer has some advantages over the Shift-Register observer proposed in Fornasini and Valcher (2013).

- The Luenberger-like observer evaluates state estimate $\hat{x}_s(t)$ only based on $\hat{x}_s(t-1)$, $y(t)$ and $u(t-1)$ instead of the complete knowledge of the input and output trajectory $\{y(t-r_{min}), u(t-r_{min}), y(t-r_{min}+1), u(t-r_{min}+1), \cdots, y(t), u(t)\}$. Hence, it requires lower computational effort.
- The Luenberger-like observer delivers state estimate during transient phase and may provide the correct state estimate earlier.

Now, the Luenberger-like observer is compared with the Multiple States observer introduced also in Fornasini and Valcher (2013). The basic idea behind the Multiple States observer is to apply a bank of observers in parallel. The number of observers is equal to the maximal number of states which can not be distinguished directly based on output. The state estimate of each observer is initialized as one state compatible with the output at initial time. Each observer updates the system state based on the state equation (2.32). During estimation procedure, if the state estimate of observers is not compatible with the current output, then it is reinitialized as one state that can generate the current output. Therefore, compared with Multiple States observers, the Luenberger-like observer (6.70) reduces computational effort. In some cases the required computational effort is reduced even by factor $2^n - 1$ which is the maximal number of states that can not be distinguished directly based on output.

4.2 Unknown Input Observer

In this part, unknown input observers for BCNs with unknown inputs (3.31)–(3.32) will be introduced. If the conditions (a) and (b) in Theorem 3.19 are satisfied and the state estimate can be decoupled from the unknown inputs, then an unknown input observer can be designed. The state observer perfectly decouples the state estimate from the effect of the unknown inputs.

Based on the Luenberger-like observer given in Theorem 4.3, an approach to design unknown input observers for the BCN with unknown inputs (3.31)–(3.32) is proposed. Applying the matrices $\tilde{L}$ and $\tilde{H}$ given by (3.45), i.e., $\tilde{L} = L\mathbf{1}_{2^b}$ and $\tilde{H} = H\mathbf{1}_{2^q}$, an unknown input observer can be constructed as

$$\begin{aligned} \hat{x}(0) &= \tilde{H}^{\mathrm{T}} y(0), \\ \hat{x}(t) &= \Phi_{2^n}^{\mathrm{T}}(I_{2^n} \otimes \tilde{H}^{\mathrm{T}})\tilde{L}\hat{x}(t-1)u(t-1)y(t), t \geq 1, \end{aligned} \tag{4.9}$$

where $\hat{x} \in \mathbb{R}^{2^n \times 1}$ is the state estimate. The state $\delta_{2^n}^k$ is a candidate for the state of the BCN with unknown inputs (3.31)–(3.32) at time t, only if the k-th entry in the vector $\hat{x}$ is non-zero, i.e. $[\hat{x}(t)]_k \neq 0$.

The convergence of the unknown input observer (4.9) is shown in Theorem 4.6.

Theorem 4.6 *Given a BCN with unknown inputs (3.31)–(3.32) and assume that the minimal decouplability index is r'_{min}. Then the state observer* (4.9) *can always provide the state of the BCN without estimation error, i.e. $\hat{x}(t) = x(t)$, for any time $t \geq r'_{min}$, no matter what the unknown inputs $\xi(t)$ and $\omega(t)$ are.*

Proof. As the state estimate of the unknown input observer (4.9) $\hat{x}(0)$ at time 0 is initialized as $\tilde{H}^{\mathrm{T}} y(0)$ and $y(0)$ belongs to the set Δ_{2^p}, it is easy to see that $\hat{x}(0)$ is one column vector of the matrix $\Omega_0 = \tilde{H}^{\mathrm{T}}$. If $\hat{x}(0) = \tilde{H}^{\mathrm{T}} y(0) \in \Delta_{2^n}$, then the output $y(0)$ can already determine the state correctly. Furthermore, according to condition (b) in Theorem 3.19, $\hat{x}(t) = x(t)$ is obtained for $t > 0$. Otherwise the vector $\hat{x}(0)$ contains more than one non-zero entry, i.e. one column vector of the matrix Θ_0. Due to $u(t) \in \Delta_{2^m}$ and $y(t) \in \Delta_{2^p}$, the vector $\hat{x}(0)u(0)y(1)$ is a column vector of the matrix $\Theta_0 \otimes I_{2^{m+p}}$. In this case, according to (3.46) and (4.9), $\hat{x}(1)$ is a column vector of the matrix Ω_1 calculated by (3.46). Repeating the same procedure until time r'_{min}, it can be seen that the vector $\hat{x}(t), t = 0, 1, \cdots, r'_{min}$ is one column vector of the matrix Ω_t. In addition, based on condition (a) in Theorem 3.19, all the columns of the matrix $\Omega_{r'_{min}}$ contain only at most one non-zero entry. Hence, $\hat{x}(r'_{min}) \in \Delta_{2^n}$ holds, i.e. the state estimate is equal to the correct state. Again, according to condition (b) in Theorem 3.19 there is $\hat{x}(t) = x(t)$ for any time $t \geq r'_{min}$. □

As mentioned before, the state $\delta_{2^n}^k$ is a candidate of state at time t, only if $[\hat{x}(t)]_k \neq 0$. The vector $\hat{x}(t)$ represents the set of candidate states that are compatible with the input and output trajectory $\{y(0), u(0), \cdots, y(t), u(t)\}$. As the real state $x(t)$ is compatible with the input and output trajectory, the state estimate $\hat{x}(t)$ provided by the unknown input observer (4.9) represents a set of states containing the real state. This is because of the way how the observer converges. At time $t+1$ the states are removed from the set if they are not consistent with the system equations (3.31)–(3.32). The real state corresponds to a consistent trajectory, so it is always part of the estimated set represented by $\hat{x}(t)$.

4.2.1 Unknown Input Estimation

In this part, it is shown that, based on the state estimate provided by the unknown input observer (4.9), the unknown inputs $\xi(t)$ and $\omega(t)$ can be estimated.

Consider the BCN with unknown inputs (3.31)–(3.32). Applying Proposition 2.4, (3.31) can be equivalently written as

$$x(t+1) = LW_{[2^{m+n},2^b]}x(t)u(t)\xi(t). \tag{4.10}$$

If the state estimates $\hat{x}(t)$, $\hat{x}(t+1)$ are provided by the unknown input observer (4.9) and the input $u(t)$ is given, then all possible unknown inputs $\xi(t)$ that satisfy (4.10) can be directly determined by

$$\hat{\xi}(t) = \left(LW_{[2^{m+n},2^b]}\hat{x}(t)u(t)\right)^{\mathrm{T}}\hat{x}(t+1) \tag{4.11}$$

where $\delta_{2^b}^k$ is a candidate of unknown input $\xi(t)$, only if $[\hat{\xi}(t)]_k \neq 0$.

In order to estimate unknown input $\omega(t)$, one needs to consider (3.32). Applying Proposition 2.4, (3.32) can be equivalently written as

$$y(t) = HW_{[2^n,2^q]}x(t)\omega(t). \tag{4.12}$$

Based on the state estimate $\hat{x}(t)$ and the output $y(t)$, the unknown input $\omega(t)$ can be determined by

$$\hat{\omega}(t) = \left(HW_{[2^n,2^q]}\hat{x}(t)\right)^{\mathrm{T}}y(t) \tag{4.13}$$

where $\delta_{2^q}^k$ is a candidate of unknown input $\omega(t)$, only if $[\hat{\omega}(t)]_k \neq 0$.

From the results obtained above, the procedure to estimate the unknown inputs $\xi(t)$ and $\omega(t)$ based on the state estimate provided by the unknown input observer (4.9) can be summarized as follows:

$$\hat{x}(0) = \tilde{H}^{\mathrm{T}} y(0), \tag{4.14}$$

$$\hat{x}(t+1) = \Phi_{2^n}^{\mathrm{T}} (I_{2^n} \otimes \tilde{H}^{\mathrm{T}}) \tilde{L} \hat{x}(t) u(t) y(t+1), \tag{4.15}$$

$$\hat{\xi}(t) = \left(L W_{[2^{m+n}, 2^b]} \hat{x}(t) u(t)\right)^{\mathrm{T}} \hat{x}(t+1), \tag{4.16}$$

$$\hat{\omega}(t) = \left(H W_{[2^n, 2^q]} \hat{x}(t)\right)^{\mathrm{T}} y(t), \; t \geq 1, \tag{4.17}$$

where the matrices $\tilde{L}$ and $\tilde{H}$ are given by (3.45).

The vector $\hat{\xi}(t)$ and $\hat{\omega}(t)$ correspond, respectively, to the set of all unknown inputs ξ and ω that can explain the relationship between $\hat{x}(t)$ and $\hat{x}(t+1)$ and the relationship between $\hat{x}(t)$ and $y(t)$. Note that the real states $x(t)$ and $x(t+1)$ belong to the sets represented by $\hat{x}(t)$ and $\hat{x}(t+1)$, respectively. Hence, the true unknown inputs $\xi(t)$ and $\omega(t)$ belong to the sets of candidate unknown inputs represented by $\hat{\xi}(t)$ and $\hat{\omega}(t)$ for any time $t \geq 0$.

Remark 4.7 As a special case, the proposed approaches can also be applied to the BCN with only unknown inputs in the state equation (3.31) (i.e. the BCN described by (3.31) and (2.33)). For this purpose, the matrices $\tilde{L}$ and $\tilde{H}$ are determined by $\tilde{L} = L\mathbf{1}_{2^b}$, $\tilde{H} = H$. To handle the BCN with only unknown inputs in the output equation (3.32) (i.e. the BCN described by (2.32) and (3.32)), the matrices $\tilde{L}$ and $\tilde{H}$ should be set to $\tilde{L} = L$, $\tilde{H} = H\mathbf{1}_{2^q}$.

4.2.2 Example

Consider the following BCN with unknown inputs $\xi(t)$ and $\omega(t)$

$$\begin{cases} X_1(t+1) = U(t) \wedge \neg X_6(t), \\ X_2(t+1) = \neg X_1(t), \\ X_3(t+1) = \neg X_1(t) \wedge (X_5(t) \vee X_3(t)), \\ X_4(t+1) = X_2(t) \wedge \neg X_6(t), \\ X_5(t+1) = (X_4(t) \vee \neg X_3(t)) \wedge \xi(t), \\ X_6(t+1) = X_5(t) \wedge (\neg X_6(t) \vee \neg X_2(t)), \\ Y_1(t) = X_2(t) \vee \omega(t), \\ Y_2(t) = X_3(t), \\ Y_3(t) = X_5(t). \end{cases} \tag{4.18}$$

This BCN with unknown inputs $\xi(t)$ and $\omega(t)$ can be described by the algebraic form (3.31)–(3.32).

$$\begin{cases} x(t+1) = L\xi(t)x(t)u(t), \\ y(t) = H\omega(t)x(t), \end{cases} \tag{4.19}$$

where $L \in \mathcal{L}_{64\times 256}$, $H \in \mathcal{L}_{8\times 128}$. As already shown in Section 3.3.3, the BCN with unknown inputs (4.19) is unknown input decouplable and the minimal decouplability index is 5. Assume that the input and output data are as given in Table 4.1.

Table 4.1 Input and output data (logical value)

t	0	1	2	3	4	5	6	7	8	9
U	1	1	1	1	1	0	1	0	1	
ξ	1	1	1	1	0	1	1	0	1	
ω	1	1	0	0	0	0	0	0	0	0
Y_1	1	1	1	0	0	1	1	1	0	1
Y_2	1	0	0	0	0	1	1	1	0	0
Y_3	0	0	1	1	1	0	0	0	0	1

As the outputs $y_2 = x_3$ and $y_3 = x_5$ are not influenced by the unknown input ω, the performance of the state observer will be shown only with the help of the state $x_1x_2x_4x_6 \in \Delta_{16}$.

It can be seen in Fig. 4.1 that the state estimate provided by the unknown input observer (4.9) converges to the correct one at time $t = 4$.

Furthermore, the unknown inputs $\xi(t)$ and $\omega(t)$ are estimated by (4.16)–(4.17). As shown in Fig. 4.2, at time $t = 1, 2, 3, 4, 7, 8$ the unknown input $\xi(t)$ is correctly estimated. At time $t = 0, 5, 6$ the unknown input $\xi(t)$ can not be uniquely estimated, but the true unknown input $\xi(t)$ is included in the set of candidates represented by $\hat{\xi}(t)$. For the unknown input ω, the correct estimation is provided at time $t = 1, 3, 4, 8$. At time $t = 0, 2, 5, 6, 7, 9$, though the unknown input $\omega(t)$ can not be correctly estimated, the real unknown input $\omega(t)$ belongs to the set of candidates represented by $\hat{\omega}(t)$.

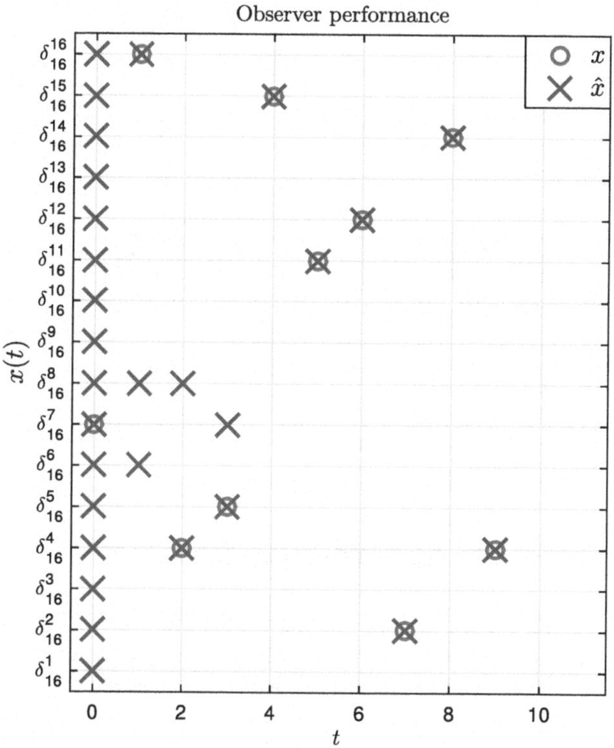

Figure 4.1 State estimate delivered by the unknown input observer (4.9) (x: real state, $\hat{x}$: state estimate)

4.3 Reduced-order Observer

The aim of this section is to introduce an approach to design reduced-order observers. It will be shown that using the reduced-order observer for state estimation of BCNs, less computational effort is required than applying the Luenberger-like observer (4.4).

In order to construct the reduced-order observer later, it is necessary to propose an approach to determine reducible state variables. At first, the concept of reducible state variables will be introduced. After that, an approach to find output variables to determine reducible state variables will be proposed. Based on this, transformation matrices for state and output coordinate transformations will be given.

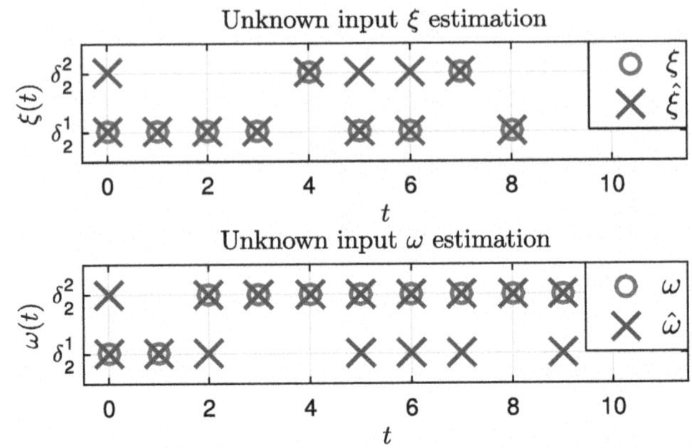

Figure 4.2 Estimate of the unknown inputs ξ and ω (ξ and ω: real unknown input, $\hat{\xi}$ and $\hat{\omega}$: estimate of unknown input)

4.3.1 Concept of Reducible State Variables

Usually, an observer constructs a state estimate of the entire internal states. However, a part of the states in linear systems can already be directly inferred from the system outputs. The elimination of redundancy provides one possible way to construct a reduced-order observer (Luenberger, 1971). Based on this idea, it is convenient to introduce the concept of reducible state variables of BCNs.

Definition 4.8 [Reducible State Variables] Considering a BCN (2.34)–(2.35), the τ state variables $\{X_{i_1}, X_{i_2}, \cdots, X_{i_\tau}\}$ are said to be reducible if there is a subset of output variables $\{Y_{j_1}, Y_{j_2}, \cdots, Y_{j_\tau}\}$ so that the knowledge of these output variables $\{Y_{j_1}(t), Y_{j_2}(t), \cdots, Y_{j_\tau}(t)\}$ at each time instant is enough to determine the state variables $\{X_{i_1}(t), X_{i_2}(t), \cdots, X_{i_\tau}(t)\}$ uniquely.

It is important to note that if state variables $X_{i_1}, X_{i_2}, \cdots, X_{i_\tau}$ are reducible state variables with respect to the set of output variables $\Omega = \{Y_{j_1}, Y_{j_2}, \cdots, Y_{j_\tau}\}$, then the state variables $X_{i_1}, X_{i_2}, \cdots, X_{i_\tau}$ are also reducible state variables with respect to any set of output variables satisfying $\bar{\Omega} \supset \Omega$. However, the knowledge of output variables in the relative complement of Ω with respect to $\bar{\Omega}$, denoted by $\bar{\Omega} \backslash \Omega$, do not provide any additional information. Therefore, the main purpose of this section is to determine the maximum number of reducible state variables corresponding to the minimal number of output variables.

At first the definition of coordinate transformation of BCNs introduced by Cheng et al. (see Chapter 10 in Cheng et al. (2011)) is considered.

Definition 4.9 [Coordinate Transformation of BCNs (Cheng et al., 2011)] Consider two column vectors $Z = [Z_1\ Z_2\ \cdots\ Z_n]^{\mathrm{T}}$ and $X = [X_1\ X_2\ \cdots\ X_n]^{\mathrm{T}}$. The mapping $G : \mathcal{D}^n \to \mathcal{D}^n$ defined by $X = [X_1\ X_2\ \cdots\ X_n]^{\mathrm{T}} \mapsto Z = [Z_1\ Z_2\ \cdots\ Z_n]^{\mathrm{T}}$ is called a coordinate transformation if G is one-to-one function and vice versa.

If two BCNs are related by a coordinate transformation, then the resulting BCN still realizes the same input-output mapping and the two BCNs are equivalent (Cheng et al., 2011).

Let z_i and x_j be, respectively, the vector form of the variables Z_i and X_j. Denoting $z = \ltimes_{i=1}^{n} z_i \in \Delta_{2^n}$ and $x = \ltimes_{j=1}^{n} x_j \in \Delta_{2^n}$, the mapping G can be described in Cheng et al. (2011) by

$$z = T_G x \tag{4.20}$$

where the matrix $T_G \in \mathcal{L}_{2^n \times 2^n}$ is nonsingular and orthogonal. Recall that each column of a logical matrix contains exactly one non-zero entry. Due to this special structure, the nonsingular logical matrix T_G is a permutation matrix. Because of the mapping G it is clear that once state z is uniquely determined, state x is also known and vice versa. Since the output variables to determine reducible state variables are not limited to the first τ output variables, i.e., $\{Y_1, Y_2, \cdots, Y_\tau\}$, the output transformation can also be applied to determine reducible state variables. However, different from the state transformation, not all permutation matrices can be selected as transformation matrix for the output coordinate transformation. The reason for this is that after performing the transformation on the output, diverse physical, biological or biochemical meanings of the variables $Y_1, Y_2, \cdots, Y_p$ shall remain in the new output $\tilde{y}$. Accordingly $\tilde{y} = \ltimes_{k=1}^{p} y_{j_k}$ should hold. Otherwise, the measured output y needs to be dealt with at each time instant it is received. Due to this, additional computational effort is required. For the sake of clarity, let the transformed output be partitioned as $\tilde{y} = \tilde{y}_1 \tilde{y}_2$ with $\tilde{y}_1 = \ltimes_{k=1}^{\tau} y_{j_k}$. To show the relationship between $\tilde{y}$ and y, it is helpful to recall the definition of σ-swap matrix W_σ.

Definition 4.10 (Wang et al. (2016)) Let σ be a permutation of the set $\{1, 2, \cdots, p\}$. There is a unique matrix W_σ called σ-swap matrix, satisfying

$$\tilde{y} = \ltimes_{k=1}^{p} y_{[\sigma]_k} = W_\sigma \ltimes_{k=1}^{p} y_k = W_\sigma y. \tag{4.21}$$

Note that the σ-swap matrix W_σ is a more general swap matrix than the matrix given in Proposition 2.4. Besides, W_σ is a special permutation matrix and can be represented by the matrix multiplication of swap matrices. For instance, if $p = 4$ and $\sigma = [2\ 4\ 1\ 3]$, then the corresponding σ-swap matrix W_σ can be expressed by $W_\sigma = (I_2 \otimes W_{[4,2]})W_{[2,2]}$.

The matrix T_G satisfying (4.20) is orthogonal as its transpose is equal to its inverse (i.e. $T_G^{-1} = T_G^{\mathrm{T}}$). Applying the coordinate transformation on state space (4.20) and the output transformation (4.21), the BCN (2.32)–(2.33) becomes

$$z(t+1) = \tilde{L}z(t)u(t), \tag{4.22}$$

$$\tilde{y}(t) = \tilde{H}z(t) \tag{4.23}$$

where $\tilde{L} = T_G L_{eq} T_G^{\mathrm{T}}$ and $\tilde{H} = W_\sigma H T_G^{\mathrm{T}}$.

Without loss of generality, let the transformed reducible state variables be denoted by $\{Z_1, Z_2, \cdots, Z_\tau\}$. Otherwise, one can always transform the vector form of reducible state variables $\{X_{i_1}, X_{i_2}, \cdots, X_{i_\tau}\}$ back to the vector form of $\{Z_1, Z_2, \cdots, Z_\tau\}$ by using an σ-swap matrix W_σ. Let $\tilde{z}_1(t) = \ltimes_{i=1}^{\tau} z_i(t)$ and $\tilde{z}_2(t) = \ltimes_{i=\tau+1}^{n} z_i(t)$. One can obtain that $\tilde{z}_1(t)\tilde{z}_2(t) = z(t)$. Our purpose is to introduce an approach to find the maximum number of reducible state variables $Z_1, Z_2, \cdots, Z_\tau$ and the corresponding transformation matrices T_G and W_σ so that the BCN in the form of (4.22)–(4.23) can be rewritten as

$$\tilde{z}_1(t+1)\tilde{z}_2(t+1) = \tilde{L}\tilde{z}_1(t)\tilde{z}_2(t)u(t), \tag{4.24}$$

$$\tilde{y}_1(t) = \tilde{z}_1(t), \tag{4.25}$$

$$\tilde{y}_2(t) = \bar{H}\tilde{z}_1(t)\tilde{z}_2(t), \tag{4.26}$$

where $\tilde{y}(t) = \tilde{y}_1(t)\tilde{y}_2(t)$ is the output satisfying (4.21) and $\tilde{z}_1 \in \Delta_{2^\tau}$ is the vector expression of τ reducible state variables.

It is worth pointing out that (4.25) can be generally replaced by $\tilde{y}_1(t) = P\tilde{z}_1(t)$ where P is a permutation matrix of the dimensions $2^\tau \times 2^\tau$. However, if $\tilde{y}_1^* = P^{\mathrm{T}}\tilde{y}_1$ and $\tilde{y}_2^* = \tilde{y}_2$ hold, then there is

$$\tilde{y}_1^* = P^{\mathrm{T}}\tilde{y}_1 = P^{\mathrm{T}}P\tilde{z}_1(t) = \tilde{z}_1(t)$$

which actually has the same form as (4.25).

Multiplying (4.25) with (4.26) together and recalling Proposition 2.5 and Proposition 2.7, one gets that

$$\begin{aligned}\tilde{y}(t) = \tilde{y}_1(t)\tilde{y}_2(t) = \tilde{z}_1(t)\bar{H}\tilde{z}_1(t)\tilde{z}_2(t) \\ &= (I_{2^\tau} \otimes \bar{H})\tilde{z}_1(t)\tilde{z}_1(t)\tilde{z}_2(t) \\ &= (I_{2^\tau} \otimes \bar{H})\Phi_{2^\tau}\tilde{z}_1(t)\tilde{z}_2(t) \\ &= (I_{2^\tau} \otimes \bar{H})\Phi_{2^\tau}\tilde{z}(t).\end{aligned} \tag{4.27}$$

After comparing (4.27) to (4.23), one shall find a matrix $\bar{H} \in \mathcal{L}_{2^{p-\tau} \times 2^n}$ with the largest possible τ and the transformation matrices T_G and W_σ so that

$$(I_{2^\tau} \otimes \bar{H})\Phi_{2^\tau} = \tilde{H} = W_\sigma H T_G^{\mathrm{T}}. \tag{4.28}$$

For the sake of clarity, recall the Kronecker product of matrices. It can be got that

$$I_{2^\tau} \otimes \bar{H} = \mathrm{diag}(\underbrace{\bar{H}, \bar{H}, \cdots, \bar{H}}_{2^\tau}) = \underbrace{\begin{bmatrix} \bar{H} & 0 & \cdots & 0 \\ 0 & \bar{H} & \cdots & 0 \\ \vdots & \vdots & \ddots & \vdots \\ 0 & 0 & \cdots & \bar{H} \end{bmatrix}}_{2^\tau}.$$

Multiplying $I_{2^\tau} \otimes \bar{H}$ with the power-reducing matrix Φ_{2^τ}, there is

$$(I_{2^\tau} \otimes \bar{H})\Phi_{2^\tau} = \mathrm{diag}(\bar{H}\delta_{2^\tau}^1, \bar{H}\delta_{2^\tau}^2, \cdots, \bar{H}\delta_{2^\tau}^{2^\tau}).$$

Partition the matrix $\bar{H}$ into 2^τ blocks as

$$\bar{H} = [\bar{H}_1 \ \ \bar{H}_2 \ \ \cdots \ \ \bar{H}_{2^\tau}], \text{ with } \bar{H}_i \in \mathcal{L}_{2^{p-\tau} \times 2^{n-\tau}}. \tag{4.29}$$

The matrix $\tilde{H}$ should have the following structure:

$$\tilde{H} = \begin{bmatrix} \bar{H}_1 & 0 & \cdots & 0 \\ 0 & \bar{H}_2 & \cdots & 0 \\ \vdots & \vdots & \ddots & \vdots \\ 0 & 0 & \cdots & \bar{H}_{2^\tau} \end{bmatrix}. \tag{4.30}$$

Based on (4.30), a necessary condition for the existence of reducible state variables is given as follows.

Lemma 4.11 If there is a row in the matrix H containing more than 2^{n-1} non-zero entries, then it is impossible to find reducible state variables of the BCN (2.34)–(2.35).

Proof. Note that W_σ and T_G^{T} are permutation matrices. Multiplying any logical matrix with the matrices W_σ and T_G^{T}, the maximum number of non-zero entries in one row of the matrix does not change. According to (4.30), the maximum number of non-zero entries in a row of matrix $\tilde{H}$ is $2^{n-\tau}$. If one row of the matrix H contains more than 2^{n-1} non-zero entries, then τ is 0 which implies that no reducible state variable exists. □

Remark 4.12 Lemma 4.11 also gives a necessary condition for the existence of the reduced-order observer of BCNs.

4.3.2 Conditions on The Transformation Matrices W_σ and T_G

After having introduced the concept of reducible state variables, in this subsection conditions on the transformation matrices will be studied in order to transform the matrix H into a matrix with a structure similar to (4.30). Based on the conditions given in this section, a computationally more efficient algorithm than the exhaustive search algorithm will be given later in Section 4.3.3 to find the transformation matrices W_σ and T_G.

For this purpose, some conditions that W_σ and T_G should satisfy are given. Let σ and $\bar{\sigma}$ be two permutations of the set $\{1, 2, \cdots, p\}$ with $[\bar{\sigma}]_i = [\sigma]_i, i = \tau+1, \tau+2, \cdots, p$. Assume that there is a transformation matrix T_G so that together with the σ-swap matrix W_σ corresponding to σ, the matrix H is transformed into a block diagonal matrix (i.e. $W_\sigma H T_G^{\mathrm{T}}$) with a structure similar to (4.30). Then the following result can be obtained.

Theorem 4.13 *Let $W_{\bar{\sigma}}$ be the σ-swap matrix corresponding to the sequence $\bar{\sigma}$. The matrix $W_{\bar{\sigma}} W_\sigma H T_G^{\mathrm{T}} W_{\bar{\sigma}}^{\mathrm{T}}$ is a block diagonal matrix with a structure similar to* (4.30).

Proof. Assume that $\tilde{H} = W_\sigma H T_G^{\mathrm{T}} = \mathrm{diag}(\bar{H}_1, \bar{H}_2, \cdots, \bar{H}_{2^\tau})$. Since the last $p-\tau$ entries in the sequences $\bar{\sigma}$ and σ are the same, the matrix $W_{\bar{\sigma}}$ can be generally expressed as

$$W_{\bar{\sigma}} = M \otimes I_{2^{p-\tau}} \tag{4.31}$$

where M is a permutation matrix of the dimensions $2^{\tau} \times 2^{\tau}$. Partition the matrix $\tilde{H}$ into $2^{2\tau}$ blocks as follows

$$\tilde{H} = \begin{bmatrix} \tilde{H}_{11} & \tilde{H}_{12} & \cdots & \tilde{H}_{12^{\tau}} \\ \tilde{H}_{21} & \tilde{H}_{22} & \cdots & \tilde{H}_{22^{\tau}} \\ \vdots & \vdots & \ddots & \vdots \\ \tilde{H}_{2^{\tau}1} & \tilde{H}_{2^{\tau}2} & \cdots & \tilde{H}_{2^{\tau}2^{\tau}} \end{bmatrix} \tag{4.32}$$

where each block $\tilde{H}_{ij}$, $i = 1, 2, \cdots, 2^{\tau}$; $j = 1, 2, \cdots, 2^{\tau}$ is of the dimensions $2^{p-\tau} \times 2^{n-\tau}$. Then, it can be obtained that

$$\tilde{H}_{ij} = \begin{cases} \bar{H}_i, & \text{if } i = j, \\ \mathbf{0}_{2^{p-\tau}} \otimes \mathbf{0}^{\mathrm{T}}_{2^{n-\tau}}, & \text{if } i \neq j. \end{cases} \tag{4.33}$$

According to Definition 2.1, it holds that

$$\begin{aligned} W_{\tilde{\sigma}} \tilde{H} W_{\tilde{\sigma}}^{\mathrm{T}} &= (M \otimes I_{2^{p-\tau}}) \tilde{H} (M^{\mathrm{T}} \otimes I_{2^{p-\tau}}) \\ &= M \tilde{H} (M^{\mathrm{T}} \otimes I_{2^{p-\tau}} \otimes I_{2^{n-p}}) \\ &= M \tilde{H} (M^{\mathrm{T}} \otimes I_{2^{n-\tau}}) = M \tilde{H} M^{\mathrm{T}}. \end{aligned}$$

Notice that $\tau \leq p$ and 2^{τ} is a factor of 2^{p}. If the matrix $W_{\tilde{\sigma}} \tilde{H} W_{\tilde{\sigma}}^{\mathrm{T}}$ is partitioned in a similar way as (4.32), then the (i, j)-th block of the matrix $W_{\tilde{\sigma}} \tilde{H} W_{\tilde{\sigma}}^{\mathrm{T}}$ is calculated by

$$\sum_{t=1}^{2^{\tau}} M_{j,t} \cdot \left(\sum_{k=1}^{2^{\tau}} M_{i,k} \cdot \tilde{H}_{kt} \right) = \sum_{t=1}^{2^{\tau}} \left(\sum_{k=1}^{2^{\tau}} M_{j,t} \cdot M_{i,k} \cdot \tilde{H}_{kt} \right). \tag{4.34}$$

According to (4.33), $\tilde{H}_{kt}$ is a non-zero matrix only if $k = t$. Therefore, (4.34) can be simplified as

$$\sum_{t=1}^{2^{\tau}} M_{j,t} \cdot \left(\sum_{k=1}^{2^{\tau}} M_{i,k} \cdot \tilde{H}_{kt} \right) = \sum_{t=1}^{2^{\tau}} M_{j,t} \cdot M_{i,t} \cdot \tilde{H}_{tt}.$$

Recall that each row in matrix M contains only one entry of 1. The (i, j)-th block of the matrix $W_{\tilde{\sigma}} \tilde{H} W_{\tilde{\sigma}}^{\mathrm{T}}$ is a non-zero matrix if and only if $i = j$. That means the matrix $W_{\tilde{\sigma}} \tilde{H} W_{\tilde{\sigma}}^{\mathrm{T}}$ has a structure similar to (4.30). □

Theorem 4.13 implies that the order of the output variables $\{Y_{j_1}, Y_{j_2}, \cdots, Y_{j_\tau}\}$ selected to determine reducible state variables does not play any role.

Next, a necessary and sufficient condition that the transformation matrices W_σ and T_G should satisfy, will be given so that an output variable Y_{j^*} can be selected as one candidate output variable.

Theorem 4.14 *An output variable Y_{j^*} can be selected to determine reducible state variables if and only if a permutation matrix $T_G \in \mathbb{R}^{2^n \times 2^n}$ can be found so that*

$$(I_2 \otimes \mathbf{1}^{\mathrm{T}}_{2^{p-1}})W_\sigma H T_G^{\mathrm{T}} = I_2 \otimes \mathbf{1}^{\mathrm{T}}_{2^{n-1}}. \tag{4.35}$$

Proof. (Sufficiency) Assume that (4.35) holds. Partition the logical matrix $\tilde{H} = W_\sigma H T_G^{\mathrm{T}}$ as

$$W_\sigma H T_G^{\mathrm{T}} = \begin{bmatrix} \tilde{H}_{11} & \tilde{H}_{12} \\ \tilde{H}_{21} & \tilde{H}_{22} \end{bmatrix} \tag{4.36}$$

where the matrices $\tilde{H}_{11}, \tilde{H}_{12}, \tilde{H}_{21}, \tilde{H}_{22}$ have the dimension $2^{p-1} \times 2^{n-1}$. According to the condition (4.35), it is clear that

$$\begin{bmatrix} \mathbf{1}^{\mathrm{T}}_{2^{p-1}}\tilde{H}_{11} & \mathbf{1}^{\mathrm{T}}_{2^{p-1}}\tilde{H}_{12} \\ \mathbf{1}^{\mathrm{T}}_{2^{p-1}}\tilde{H}_{21} & \mathbf{1}^{\mathrm{T}}_{2^{p-1}}\tilde{H}_{22} \end{bmatrix} = I_2 \otimes \mathbf{1}^{\mathrm{T}}_{2^{n-1}} = \begin{bmatrix} \mathbf{1}^{\mathrm{T}}_{2^{n-1}} & \mathbf{0}^{\mathrm{T}}_{2^{n-1}} \\ \mathbf{0}^{\mathrm{T}}_{2^{n-1}} & \mathbf{1}^{\mathrm{T}}_{2^{n-1}} \end{bmatrix} \tag{4.37}$$

which indicates that

$$\begin{cases} \tilde{H}_{11} & \in \mathcal{L}_{2^{p-1} \times 2^{n-1}}, \\ \tilde{H}_{12} & = \mathbf{0}_{2^p-1} \otimes \mathbf{0}^{\mathrm{T}}_{2^n-1}, \\ \tilde{H}_{21} & = \mathbf{0}_{2^p-1} \otimes \mathbf{0}^{\mathrm{T}}_{2^n-1}, \\ \tilde{H}_{22} & \in \mathcal{L}_{2^{p-1} \times 2^{n-1}}. \end{cases} \tag{4.38}$$

Hence, the matrix $W_\sigma H T_G^{\mathrm{T}}$ is a block diagonal matrix with the same structure as (4.30) and the output variable Y_{j^*} can be selected as a candidate.

(Necessity) Assume that the output variable Y_{j^*} can be selected for determining reducible state variables. According to Theorem 4.13, the matrix $W_\sigma H T_G^{\mathrm{T}}$ is a block matrix with the structure shown in (4.30). According to (4.28), there is a logical matrix $\bar{H} \in \mathcal{L}_{2^{p-\tau} \times 2^n}$ so that $(I_{2^\tau} \otimes \bar{H})\Phi_\tau = W_\sigma H T_G^{\mathrm{T}}$. Based on this and left multiplying $I_2 \otimes \mathbf{1}^{\mathrm{T}}_{2^{p-1}}$, one gets

$$(I_2 \otimes \mathbf{1}^{\mathrm{T}}_{2^{p-1}})W_\sigma H T_G^{\mathrm{T}} = (I_2 \otimes \mathbf{1}^{\mathrm{T}}_{2^{p-1}})(I_{2^\tau} \otimes \bar{H})\Phi_{2^\tau}. \tag{4.39}$$

Through a simple calculation, the dimensions of the matrices $I_2 \otimes \mathbf{1}^{\mathrm{T}}_{2^{\tau-1}} \in \mathbb{R}^{2\times 2^{\tau}}$, $\mathbf{1}^{\mathrm{T}}_{2^{p-\tau}} \in \mathbb{R}^{1\times 2^{p-\tau}}$, $I_{2^{\tau}} \in \mathbb{R}^{1\times 2^{r}}$ and $\bar{H} \in \mathbb{R}^{2^{p-\tau}\times 2^{n}}$ are obtained. Applying the mixed-product property of the Kronecker product (3.14), (4.39) is equivalently rewritten as

$$\begin{aligned}
(I_2 \otimes \mathbf{1}^{\mathrm{T}}_{2^{p-1}})(I_{2^{\tau}} \otimes \bar{H})\Phi_{2^{\tau}} &= \left((I_2 \otimes \mathbf{1}^{\mathrm{T}}_{2^{\tau-1}} \otimes \mathbf{1}^{\mathrm{T}}_{2^{p-\tau}})(I_{2^{\tau}} \otimes \bar{H})\right)\Phi_{2^{\tau}} \\
&= \left(\left(\underbrace{I_2 \otimes \mathbf{1}^{\mathrm{T}}_{2^{\tau-1}}}_{A} \otimes \underbrace{\mathbf{1}^{\mathrm{T}}_{2^{p-\tau}}}_{B}\right)\cdot\left(\underbrace{I_{2^{\tau}}}_{C} \otimes \underbrace{\bar{H}}_{D}\right)\right)\Phi_{2^{\tau}} \\
&= \left(\left(\underbrace{(I_2 \otimes \mathbf{1}^{\mathrm{T}}_{2^{\tau-1}})}_{A}\cdot\underbrace{I_{2^{\tau}}}_{C}\right)\otimes\left(\underbrace{\mathbf{1}^{\mathrm{T}}_{2^{p-\tau}}}_{B}\cdot\underbrace{\bar{H}}_{D}\right)\right)\Phi_{2^{\tau}} \\
&= \left(\left((I_2 \otimes \mathbf{1}^{\mathrm{T}}_{2^{\tau-1}})\cdot I_{2^{\tau}}\right)\otimes\left(\mathbf{1}^{\mathrm{T}}_{2^{p-\tau}}\cdot\bar{H}\right)\right)\Phi_{2^{\tau}}.
\end{aligned} \tag{4.40}$$

Notice that $\mathbf{1}^{\mathrm{T}}_{2^{p-\tau}} \cdot \bar{H} = \mathbf{1}^{\mathrm{T}}_{2^{n}}$. (4.40) can be further rewritten as

$$\left((I_2 \otimes \mathbf{1}^{\mathrm{T}}_{2^{\tau-1}}) \otimes \mathbf{1}^{\mathrm{T}}_{2^{n}}\right)\Phi_{2^{\tau}} = \left(I_2 \otimes \mathbf{1}^{\mathrm{T}}_{2^{n+\tau-1}}\right)\Phi_{2^{\tau}}.$$

Recalling the power-reducing matrix Ω_r defined in Definition 2.6, there is

$$\begin{aligned}
\left(I_2 \otimes \mathbf{1}^{\mathrm{T}}_{2^{n+\tau-1}}\right)\Phi_{2^{\tau}} &= \left(I_2 \otimes \mathbf{1}^{\mathrm{T}}_{2^{n+\tau-1}}\right)\left[\delta^1_{2^{\tau}} \otimes \delta^1_{2^{\tau}} \;\; \delta^2_{2^{\tau}} \otimes \delta^2_{2^{\tau}} \;\; \cdots \;\; \delta^{2^{\tau}}_{2^{\tau}} \otimes \delta^{2^{\tau}}_{2^{\tau}}\right] \\
&= \left[\left(I_2 \otimes \mathbf{1}^{\mathrm{T}}_{2^{n+\tau-1}}\right)(\delta^1_{2^{\tau}} \otimes \delta^1_{2^{\tau}}) \;\; \cdots \;\; \left(I_2 \otimes \mathbf{1}^{\mathrm{T}}_{2^{n+\tau-1}}\right)(\delta^{2^{\tau}}_{2^{\tau}} \otimes \delta^{2^{\tau}}_{2^{\tau}})\right].
\end{aligned} \tag{4.41}$$

Denote the matrices B_1 and B_2, respectively, as $B_1 = \delta^1_2 \otimes \mathbf{1}^{\mathrm{T}}_{2^{n}}$ and $B_2 = \delta^2_2 \otimes \mathbf{1}^{\mathrm{T}}_{2^{n}}$. It can be easily understood that

$$I_2 \otimes \mathbf{1}^{\mathrm{T}}_{2^{n+\tau-1}} = \left[\underbrace{B_1\; B_1 \cdots B_1}_{2^{\tau-1}} \;\; \underbrace{B_2\; B_2 \cdots B_2}_{2^{\tau-1}}\right]. \tag{4.42}$$

Based on (4.42) and according to Definition 2.1, there is

$$(I_2 \otimes \mathbf{1}_{2^{n+\tau-1}}^{\mathrm{T}})(\delta_{2^\tau}^i \otimes \delta_{2^\tau}^i) = (I_2 \otimes \mathbf{1}_{2^{n+\tau-1}}^{\mathrm{T}})\delta_{2^\tau}^i \delta_{2^\tau}^i = \begin{cases} \delta_2^1 \otimes \mathbf{1}_{2^{n-\tau}}^{\mathrm{T}}, & \text{if } 1 \leq i \leq 2^{\tau-1}, \\ \delta_2^2 \otimes \mathbf{1}_{2^{n-\tau}}^{\mathrm{T}}, & \text{if } 2^{\tau-1}+1 \leq i \leq 2^\tau. \end{cases}$$

Hence, (4.41) is actually equal to

$$\begin{aligned} \left(I_2 \otimes \mathbf{1}_{2^{n+\tau-1}}^{\mathrm{T}}\right) \Phi_{2^\tau} &= \left[\delta_2^1 \otimes \mathbf{1}_{2^{n-\tau}}^{\mathrm{T}} \otimes \mathbf{1}_{2^{\tau-1}}^{\mathrm{T}} \quad \delta_2^2 \otimes \mathbf{1}_{2^{n-\tau}}^{\mathrm{T}} \otimes \mathbf{1}_{2^{\tau-1}}^{\mathrm{T}}\right] \\ &= \left[\delta_2^1 \otimes \mathbf{1}_{2^{n-1}}^{\mathrm{T}} \quad \delta_2^2 \otimes \mathbf{1}_{2^{n-1}}^{\mathrm{T}}\right] \\ &= I_2 \otimes \mathbf{1}_{2^{n-1}}^{\mathrm{T}} \end{aligned}$$

which reveals the result in the theorem. □

The result of Theorem 4.14 can be extended to a candidate set of output variables $\{Y_{j_1}, Y_{j_2}, \cdots, Y_{j_k}\}$. If the first k entries in a sequence of n components σ can be specified as $[\sigma]_i = j_i, i = 1, 2, \cdots, k$, then the corresponding σ-swap matrix W_σ satisfies $\ltimes_{k=1}^{p} y_{[\sigma]_k} = W_\sigma y$.

Theorem 4.15 *There is a set of output variables $\{Y_{j_1}, Y_{j_2}, \cdots, Y_{j_k}\}$ that can be selected to determine reducible state variables if and only if there is a permutation matrix $T_G \in \mathbb{R}^{2^n \times 2^n}$ so that it holds*

$$(I_{2^k} \otimes \mathbf{1}_{2^{p-k}}^{\mathrm{T}}) W_\sigma H T_G^{\mathrm{T}} = I_{2^k} \otimes \mathbf{1}_{2^{n-k}}^{\mathrm{T}}. \tag{4.43}$$

Proof. Theorem 4.15 can be proven in the same way as Theorem 4.14, so it is omitted. □

Theorem 4.15 gives a necessary and sufficient condition for selecting a set of output variables. Based on Theorem 4.15, finding the maximum number of reducible state variables corresponds to a optimization problem. The objective is to find a state transformation matrix T_G and an σ-swap matrix W_σ so that k is maximized under the constraint (4.43).

In order to add a new output variable $Y_{j_{k+1}}$ into the set of output variables $\{Y_{j_1}, Y_{j_2}, \cdots, Y_{j_k}\}$, one could indeed check the condition (4.43) for all $p-k$ possible candidates of $Y_{j_{k+1}}$. However, it might lead to a higher computational effort. Hence, Lemma 4.16 and Theorem 4.17 are introduced to reduce search space.

Lemma 4.16 If (4.43) is satisfied, then it also holds that

$$(I_{2^t} \otimes \mathbf{1}_{2^{p-t}}^{\mathrm{T}}) W_\sigma H T_G^{\mathrm{T}} = I_{2^t} \otimes \mathbf{1}_{2^{n-t}}^{\mathrm{T}}, \ \forall\, 0 \leq t \leq k. \tag{4.44}$$

Proof. By applying the mixed-product property of Kronecker product (3.14),

$$\begin{aligned}\left(I_{2^{k-1}} \otimes \mathbf{1}_2^{\mathrm{T}}\right)\left(I_{2^k} \otimes \mathbf{1}_{2^{n-k}}^{\mathrm{T}}\right) &= \left(I_{2^{k-1}} \otimes \mathbf{1}_2^{\mathrm{T}}\right) \cdot \left(I_{2^{k-1}} \otimes I_2 \otimes \mathbf{1}_{2^{n-k}}^{\mathrm{T}}\right) \\ &= \left(I_{2^{k-1}} \cdot I_{2^{k-1}}\right) \otimes \left(\mathbf{1}_2^{\mathrm{T}} \cdot \left(I_2 \otimes \mathbf{1}_{2^{n-k}}^{\mathrm{T}}\right)\right) \\ &= I_{2^{k-1}} \otimes \mathbf{1}_{2^{n-k+1}}^{\mathrm{T}}.\end{aligned} \tag{4.45}$$

is obtained. Similarly, one has

$$\left(I_{2^{k-1}} \otimes \mathbf{1}_2^{\mathrm{T}}\right)\left(I_{2^k} \otimes \mathbf{1}_{2^{p-k}}^{\mathrm{T}}\right) = I_{2^{k-1}} \otimes \mathbf{1}_{2^{p-k+1}}^{\mathrm{T}}. \tag{4.46}$$

Multiplying (4.43) on both sides from the left by $\left(I_{2^{k-1}} \otimes \mathbf{1}_2^{\mathrm{T}}\right)$, it is clear that

$$(I_{2^{k-1}} \otimes \mathbf{1}_2^{\mathrm{T}})(I_{2^k} \otimes \mathbf{1}_{2^{p-k}}^{\mathrm{T}})W_\sigma H T_G^{\mathrm{T}} = (I_{2^{k-1}} \otimes \mathbf{1}_2^{\mathrm{T}})(I_{2^k} \otimes \mathbf{1}_{2^{n-k}}^{\mathrm{T}}). \tag{4.47}$$

According to (4.45) and (4.46), (4.47) can be rewritten as

$$\left(I_{2^{k-1}} \otimes \mathbf{1}_{2^{p-k+1}}^{\mathrm{T}}\right) W_\sigma H T_G^{\mathrm{T}} = I_{2^{k-1}} \otimes \mathbf{1}_{2^{n-k+1}}^{\mathrm{T}} \tag{4.48}$$

which shows that (4.44) holds for $t = k - 1$. In a similar way, by left multiplication of $\left(I_{2^{k-2}} \otimes \mathbf{1}_2^{\mathrm{T}}\right)$ on both sides of (4.48), one gets

$$\left(I_{2^{k-2}} \otimes \mathbf{1}_{2^{p-k+2}}^{\mathrm{T}}\right) W_\sigma H T_G^{\mathrm{T}} = I_{2^{k-2}} \otimes \mathbf{1}_{2^{n-k+2}}^{\mathrm{T}} \tag{4.49}$$

which corresponds to (4.44) with $t = k - 2$. Repeating the same procedure by left multiplying $\left(I_{2^t} \otimes \mathbf{1}_2^{\mathrm{T}}\right)$, $t = k - 3, k - 4, \cdots, 1$, it can be concluded that (4.44) holds for any $0 \le t \le k$. □

Theorem 4.15 and Lemma 4.16 tell us that if a set of output variables $\Omega = \{Y_{j_1}, Y_{j_2}, \cdots, Y_{j_k}\}$ can be selected to determine reducible state variables, then the first t output variables $Y_{j_1}, Y_{j_2}, \cdots, Y_{j_t}$ in the set Ω can be selected to determine reducible state variables.

As a result, one can arrive at the following conclusion.

Theorem 4.17 *If a set of output variables* $\Omega = \{Y_{j_1}, Y_{j_2}, \cdots, Y_{j_{k-1}}, Y_{j_k}\}$ *can be selected to determine reducible state variables, then any nonempty subset* $\Omega_{sub} \subseteq \Omega$ *can also be selected to determine reducible state variables.*

Proof. According to Theorem 4.13, the matrix $W_{\bar{\sigma}} W_\sigma H T_G^{\mathrm{T}} W_{\bar{\sigma}}^{\mathrm{T}}$ with $W_{\bar{\sigma}}$ defined in (4.31) is a logical matrix and has a block diagonal structure similar to (4.30). It can be shown in a similar way as (4.36)–(4.38) that the equality

$$(I_{2^k} \otimes \mathbf{1}_{2^{p-k}}^{\mathrm{T}}) W_{\bar{\sigma}} W_\sigma H T_G^{\mathrm{T}} W_{\bar{\sigma}}^{\mathrm{T}} = I_{2^k} \otimes \mathbf{1}_{2^{n-k}}^{\mathrm{T}}$$

holds. This means that any rearrangement of the output variables $Y_{j_1}, Y_{j_2}, \cdots, Y_{j_k}$ in the set Ω does not change that the set Ω can be selected to determine reducible state variables. According to Lemma 4.16, there is

$$(I_{2^t} \otimes \mathbf{1}_{2^{p-t}}^{\mathrm{T}}) W_{\bar{\sigma}} W_\sigma H T_G^{\mathrm{T}} W_{\bar{\sigma}}^{\mathrm{T}} = I_{2^t} \otimes \mathbf{1}_{2^{n-t}}^{\mathrm{T}}, \quad \forall\, 0 \leq t \leq k.$$

This means that the set of output variables in Ω_{sub} obtained by taking out arbitrary t variables in the set Ω can be selected to determine reducible state variables. □

Theorem 4.17 implies a requirement of the new output variable to be taken into a candidate set of output variables, i.e. the new output variable should be a candidate output variable or belong to some other candidate sets.

4.3.3 Recursive Algorithm to Determine Transformation Matrices

In this subsection, a recursive algorithm to determine transformation matrices W_σ and T_G will be presented.

Recall that $\bar{H}$ is a logical matrix. According to (4.29), it is clear that all the blocks $\bar{H}_i, i = 1, 2, \cdots, 2^\tau$ are logical matrices. Hence, each block $\bar{H}_i$ always contains $2^{n-\tau}$ non-zero entries. The problem to find reducible state variables is reformulated as grouping all rows in the matrix H into $2^{p-\tau}$ sets containing the same number of rows so that the corresponding σ-swap matrix W_σ satisfies (4.21) and all rows in one set contain $2^{n-\tau}$ non-zero entries in total. Certainly, a brute-force search or an exhaustive search algorithm, i.e. enumerating all possible candidates of W_σ corresponding to $p!$ different permutations of the set $\{1, 2, \cdots, p\}$ and $2^n!$ different matrices T_G, can be applied directly to solve this problem. But from a computational viewpoint, these search algorithms may not be applicable in practice. Therefore, it is necessary and meaningful to introduce a computationally more efficient algorithm.

Consider the fact that the transformation matrix T_G could be any possible permutation matrix which indicates that the position of the columns in the matrix H can be arbitrarily changed. However, the number of non-zero entries of each row in the matrix H always remains the same. In contrast, the σ-swap matrix W_σ is a special permutation matrix according to Definition 4.10 and can rearrange the rows. For this reason, a method to determine W_σ will be introduced at first. After that, it will be shown how to determine T_G based on the matrix W_σ.

Determine the output transformation matrix W_σ

Recall that the order of output variables does not play any role. It will be shown that the output variables in the set $\{Y_{j_1}, Y_{j_2}, \cdots, Y_{j_\tau}\}$ can indeed be found one by one. Without loss of generality, let the indices $j_1, j_2, \cdots, j_\tau$ be arranged in the first τ elements in the list while the rest $p-\tau$ elements are in ascending order. The result is denoted as the permutation σ. According to (2.1) and Lemma 2.2, the corresponding σ-swap matrix W_σ can be calculated by

$$\begin{aligned} W_\sigma^{\mathrm{T}} &= W_{[2,2^{j_1-1}]}\left(I_2 \otimes W_{[2,2^{j_2-2}]}\right)\cdots\left(I_{2^{\tau-1}} \otimes W_{[2,2^{j_\tau-\tau}]}\right) \\ &= \prod_{t=1}^{\tau}\left(I_{2^{t-1}} \otimes W_{[2,2^{j_t-t}]} \otimes I_{2^{p-j_t}}\right). \end{aligned}$$

Notice that for a given index j_k, the matrix $I_{2^{k-1}} \otimes W_{[2,2^{j_k-k}]} \otimes I_{2^{p-j_k}}$ can be directly constructed. Let matrices $\widetilde{W}_k$ be defined as $\widetilde{W}_k = \prod_{t=1}^{k}\left(I_{2^{t-1}} \otimes W_{[2,2^{j_t-t}]} \otimes I_{2^{p-j_t}}\right)$. Then it can be seen that $W_\sigma = \widetilde{W}_\tau^{\mathrm{T}}$ with $\widetilde{W}_\tau$ recursively calculated by

$$\widetilde{W}_k = \widetilde{W}_{k-1}\left(I_{2^{k-1}} \otimes W_{[2,2^{j_k-k}]} \otimes I_{2^{p-j_k}}\right), k = 1, 2, \cdots, \tau. \tag{4.50}$$

By making use of this fact, a simple way to find the matrix W_σ is to apply a recursive algorithm. However, the matrix multiplication based on 4.3.3 may require high computational effort. In order to solve this problem, one can express the matrices $\widetilde{W}_{k-1}$ and $I_{2^{k-1}} \otimes W_{[2,2^{j_k-k}]} \otimes I_{2^{p-j_k}}$, respectively, as

$$\widetilde{W}_{k-1} = \delta_{2^p}[\theta_{k-1}(1)\ \theta_{k-1}(2)\ \cdots\ \theta_{k-1}(2^p)] \tag{4.51}$$

and

$$I_{2^{k-1}} \otimes W_{[2,2^{j_k-k}]} \otimes I_{2^{p-j_k}} = \delta_{2^p}[\rho_k(1)\ \rho_k(2)\ \cdots\ \rho_k(2^p)] \tag{4.52}$$

where $\theta_{k-1}(i)$ and $\rho_k(i)$ denote, respectively, the position of nonzero entry in the i-th column of $\widetilde{W}_{k-1}$ and $I_{2^{k-1}} \otimes W_{[2,2^{j_k-k}]} \otimes I_{2^{p-j_k}}$. Inserting (4.51)–(4.52) into (4.50), it can be obtained that

$$\begin{aligned}\widetilde{W}_k &= \delta_{2^p}[\theta_k(1)\ \theta_k(2)\ \cdots\ \theta_k(2^p)] \\ &= \delta_{2^p}[\theta_{k-1}(1)\ \theta_{k-1}(2)\ \cdots\ \theta_{k-1}(2^p)] \cdot \delta_{2^p}[\rho_k(1)\ \rho_k(2)\ \cdots\ \rho_k(2^p)] \\ &= \delta_{2^p}[\theta_{k-1}(\rho_k(1))\ \theta_{k-1}(\rho_k(2))\ \cdots\ \theta_{k-1}(\rho_k(2^p))]\end{aligned}$$

which shows that $\theta_k(i) = \theta_{k-1}(\rho_k(i))$. Hence, if the matrix $\widetilde{W}_0$ is initialized as $\widetilde{W}_0 = I_{2^p} = \delta_{2^p}[1\ 2\ \cdots\ 2^p] = [\theta_0(1)\ \theta_0(2)\ \cdots\ \theta_0(2^p)]$, then after recursive multiplying with the matrix $I_{2^{k-1}} \otimes W_{[2,2^{j_k-k}]} \otimes I_{2^{p-j_k}}$, $k = 1, 2, \cdots, \tau$, the i-th column of the matrix W_σ^{T} has one non-negative entry at the $\rho_1(\rho_2(\cdots(\rho_\tau(i))\cdots))$ -th position. In the following parts, this way will be used to perform multiplication of permutation matrices to reduce computational effort.

In the following, a recursive algorithm to find W_σ will be proposed based on Theorem 4.13-Theorem 4.17 in Section 4.3.2. According to the definition of the Kronecker product (2.1), (4.43) can be rewritten as

$$(I_{2^{k-1}} \otimes I_2 \otimes \mathbf{1}^{\mathrm{T}}_{2^{p-k}}) W_\sigma H T_G^{\mathrm{T}} = I_{2^{k-1}} \otimes I_2 \otimes \mathbf{1}^{\mathrm{T}}_{2^{n-k}}. \tag{4.53}$$

Similarly, for the output variables $Y_{j_1}, Y_{j_2}, \cdots, Y_{j_{k-1}}$, the matrices W_σ and T_G should also satisfy

$$(I_{2^{k-1}} \otimes \mathbf{1}^{\mathrm{T}}_{2^{p-k+1}}) W_\sigma H T_G^{\mathrm{T}} = I_{2^{k-1}} \otimes \mathbf{1}^{\mathrm{T}}_{2^{n-k+1}}. \tag{4.54}$$

The matrices $I_2 \otimes \mathbf{1}^{\mathrm{T}}_{2^{p-k}}$ and $\mathbf{1}^{\mathrm{T}}_{2^{p-k+1}}$ have, respectively, the structure

$$I_2 \otimes \mathbf{1}^{\mathrm{T}}_{2^{p-k}} = \left[\begin{array}{c|c} \underbrace{\begin{matrix}1\ 1 \cdots 1 \\ 0\ 0 \cdots 0\end{matrix}}_{2^{p-k}} & \underbrace{\begin{matrix}0\ 0 \cdots 0 \\ 1\ 1 \cdots 1\end{matrix}}_{2^{p-k}} \end{array}\right] \tag{4.55}$$

and

$$\mathbf{1}^{\mathrm{T}}_{2^{p-k+1}} = \left[\begin{array}{c|c} \underbrace{1\ 1 \cdots 1}_{2^{p-k}} & \underbrace{1\ 1 \cdots 1}_{2^{p-k}} \end{array}\right]. \tag{4.56}$$

After comparing (4.55) to (4.56), it can be seen that in order to consider the new output variable Y_{j_k} additionally, one should split the non-zero entries in a row into two groups at first. Then, according to the position of the non-zero entries in a group it should be checked, whether there are 2^{n-k} non-zero entries in the corresponding rows of the matrix $W_\sigma H T_G^{\mathrm{T}}$.

Based on this idea, the number of non-zero entries in the i-th row of the matrix H, denoted by $\upsilon(i), i = 1, 2, \cdots, 2^p$, is obtained as a first step. Then, the upper bound of the number of reducible state variables τ can be obtained as follows.

Theorem 4.18 *Given a BCN (2.34)–(2.35). Let* $\upsilon_{max} = \max_{i=1,2,\cdots,2^p} \upsilon(i)$. *The number of reducible state variables* τ *satisfies*

$$\tau \leq \tau_{max} = p - \lceil \log_2(\upsilon_{max}) \rceil \tag{4.57}$$

where $\lceil x \rceil$ *is the ceiling function, which maps* x *to the least integer greater than or equal to* x.

Proof. Because of the structure given in (4.30), there is a block $\bar{H}_i$ so that one row of $\bar{H}_i$ contains υ_{max} non-zero entries. As all the blocks $\bar{H}_i, i = 1, 2, \cdots, 2^\tau$ contain $2^{p-\tau}$ non-zero entries in total. Hence, it holds that

$$2^{p-\tau} \geq \upsilon_{max} \Leftrightarrow p - \tau \geq \log_2(\upsilon_{max}) \Leftrightarrow \tau \leq p - \log_2(\upsilon_{max}).$$

Since τ must be an integer, the equality in (4.57) is obtained. □

Now, candidate output variable Y_{j_1} will be found. The corresponding σ-swap matrix denoted by $\widetilde{W}_1^{\mathrm{T}}$ will be determined as a first step. Applying Proposition 2.4, it is clear that

$$\begin{aligned} \ltimes_{i=1}^{p} y_i &= W_{[2,2^{j_1-1}]} y_{j_1} y_1 y_2 \cdots y_{j_1-1} y_{j_1+1} \cdots y_p \\ &= I_{2^p}(I_{2^0} \otimes W_{[2,2^{j_1-1}]} \otimes I_{2^{p-j_1}}) y_{j_1} y_1 y_2 \cdots y_{j_1-1} y_{j_1+1} \cdots y_p \\ &= \underbrace{\widetilde{W}_0(I_{2^0} \otimes W_{[2,2^{j_1-1}]} \otimes I_{2^{p-j_1}})}_{\widetilde{W}_1} y_{j_1} y_1 y_2 \cdots y_{j_1-1} y_{j_1+1} \cdots y_p. \end{aligned}$$

The condition (4.35) should be checked according to Theorem 4.14 as a second step. In order to reduce computational complexity, let the matrix $\widetilde{W}_1$ be expressed as

$$\widetilde{W}_1 = \delta_{2^p}[\theta_1(1)\ \theta_1(2)\ \cdots\ \theta_1(2^p)]. \tag{4.58}$$

Applying Proposition 2.2, $(I_2 \otimes \mathbf{1}_{2^{p-1}}^{\mathrm{T}})\widetilde{W}_1^{\mathrm{T}} H T_G^{\mathrm{T}}$ can be rewritten as

$$(I_2 \otimes \mathbf{1}_{2^{p-1}}^{\mathrm{T}})\widetilde{W}_1^{\mathrm{T}} H T_G^{\mathrm{T}} = \left(\widetilde{W}_1 (I_2 \otimes \mathbf{1}_{2^{p-1}})\right)^{\mathrm{T}} H T_G^{\mathrm{T}} \tag{4.59}$$

$$= \left[\sum_{i=1}^{2^{p-1}} \delta_{2^p}^{\theta_1(i)} \mid \sum_{i=2^{p-1}+1}^{2^p} \delta_{2^p}^{\theta_1(i)} \right]^{\mathrm{T}} H T_G^{\mathrm{T}}$$

$$= \begin{bmatrix} \sum_{i=1}^{2^{p-1}} Row_{\theta_1(i)}\left(H T_G^{\mathrm{T}}\right) \\ \sum_{i=2^{p-1}+1}^{2^p} Row_{\theta_1(i)}\left(H T_G^{\mathrm{T}}\right) \end{bmatrix}. \tag{4.60}$$

As each column in the logical matrix H can be found in the set Δ_{2^p} thus $\sum_{i=1}^{2^p} v(\theta_1(i)) = 2^n$ and T_G can rearrange the columns in the matrix H, (4.59) shows that the condition (4.35) is equivalent to

$$\sum_{i=1}^{2^{p-1}} v(\theta_1(i)) = 2^{n-1}. \tag{4.61}$$

Express the matrix $I_{2^0} \otimes W_{[2,2^{j_1-1}]} \otimes I_{2^{p-j_1}}$ in the same way as (4.52). According to (4.51), it can be obtained that $\theta_1(i) = \theta_0(\rho_1(i)) = \rho_1(i), i = 1, 2, \cdots, 2^p$. As the index $\rho_1(i)$ only depends on j_1, one can select $j_1 \in \{1, 2, \cdots, 2^p\}$ one by one and check the condition (4.61). If (4.61) is satisfied, then, according to Theorem 4.14, the output variable Y_{j_1} is a candidate output variable. Note that there might be more than one candidate output variable Y_{j_1}. According to Theorem 4.17, these candidates are important to find the candidate sets of two output variables. Hence, if Y_{j_1} is a candidate output variable, then let the output variable Y_{j_1} be contained in the set denoted by Ξ_1.

After that, assume that there are α candidate output variables sets $\{Y_{j_1}, Y_{j_2}, \cdots, Y_{j_{k-1}}\}$ in the set $\Xi_{k-1} = \{\mathcal{A}_{k-1,1}, \mathcal{A}_{k-1,2}, \cdots, \mathcal{A}_{k-1,|\Xi_{k-1}|}\}$. Before finding a new output variable Y_{j_k}, it is necessary to give a condition equivalent to (4.43). For the sake of clarity, let the permutation matrix $\widetilde{W}_{k-1}$ corresponding to the candidate set $\mathcal{A}_{k-1,i}$ be expressed in the same way as (4.51). According to Theorem 4.15, the matrix $\widetilde{W}_{k-1}$ corresponding to each candidate set satisfies the condition (4.43). In the same way as (4.59), $(I_{2^{k-1}} \otimes \mathbf{1}_{2^{p-k+1}}^{\mathrm{T}})\widetilde{W}_{k-1}^{\mathrm{T}}$ can be rewritten as

$$
\begin{aligned}
(I_{2^{k-1}} \otimes \mathbf{1}_{2^{p-k+1}}^{\mathrm{T}})\widetilde{W}_{k-1}^{\mathrm{T}} &= \left(\widetilde{W}_{k-1}(I_{2^{k-1}} \otimes \mathbf{1}_{2^{p-k+1}})\right)^{\mathrm{T}} \\
&= \begin{bmatrix} \sum_{i=1}^{2^{p-k+1}} \left(\delta_{2^p}^{\theta_{k-1}(i)}\right)^{\mathrm{T}} \\ \sum_{i=2^{p-k+1}+1}^{2^{p-k+2}} \left(\delta_{2^p}^{\theta_{k-1}(i)}\right)^{\mathrm{T}} \\ \vdots \\ \sum_{i=2^p-2^{p-k+1}+1}^{2^p} \left(\delta_{2^p}^{\theta_{k-1}(i)}\right)^{\mathrm{T}} \end{bmatrix}.
\end{aligned} \tag{4.62}
$$

Noticing that $\left(\delta_{2^p}^{\theta_{k-1}(i)}\right)^{\mathrm{T}} HT_G^{\mathrm{T}} = Row_{\theta_{k-1}(i)}(HT_G^{\mathrm{T}})$, the fulfillment of the condition (4.43) means that for $q = 0, 1, \cdots, 2^{k-1} - 1$ it holds

$$
\sum_{i=q\cdot 2^{p-k+1}+1}^{(q+1)\cdot 2^{p-k+1}} v(\theta_{k-1}(i)) = 2^{n-k+1}, q = 0, 1, \cdots, 2^{k-1} - 1. \tag{4.63}
$$

Now a new output variable Y_{j_k} will be found additionally. Let the set Ξ_k be initialized as $\emptyset$. As mentioned in Subsection 4.3.2, Y_{j_k} should derive from some other candidate sets. Hence, all candidate sets in Ξ_{k-1} will be compared to each other. However, since output variables $\{Y_{j_1}, Y_{j_2}, \cdots, Y_{j_\tau}\}$ are found recursively, according to Theorem 4.17 it is enough to consider the two candidate sets where only one output variable differs, i.e. the sets $\mathcal{A}_{k-1,a}, \mathcal{A}_{k-1,b} \in \Xi_{k-1}$ are considered if

$$
|\mathcal{A}_{k-1,a} \cap \mathcal{A}_{k-1,b}| = k - 2. \tag{4.64}
$$

After finding the two candidate sets $\mathcal{A}_{k-1,a}$ and $\mathcal{A}_{k-1,b}$, it will be checked whether the output variables $\mathcal{A}_{k-1,a} \cup \mathcal{A}_{k-1,b} = \{Y_{j_1}, Y_{j_2}, \cdots, Y_{j_k}\}$ can be selected to determine reducible state variables. For this purpose, let the matrix $\widetilde{W}_k$ be represented as $\delta_{2^p}[\theta_k(1)\ \theta_k(2)\ \cdots\ \theta_k(2^p)]$. In a similar way as (4.63), it can be obtained that the corresponding condition (4.43) is equivalent to

$$
\sum_{i=q\cdot 2^{p-k}+1}^{(q+1)\cdot 2^{p-k}} v(\theta_k(i)) = 2^{n-k}, \quad q = 0, 1, \cdots, 2^k - 1. \tag{4.65}
$$

Based on the index j_k, the matrix $I_{2^{k-1}} \otimes W_{[2,2^{j_k-k}]} \otimes I_{2^{p-j_k}}$ can be uniquely constructed and is expressed in the same way as (4.52). As the matrix $\widetilde{W}_k$ is calculated by (4.50), $\theta_k(i) = \theta_{k-1}(\rho_k(i))$ can be obtained. Notice that the q-th block $\delta_{2^p}[\rho_k(q \cdot 2^{p-k+1} + 1), \rho_k(q \cdot 2^{p-k+1} + 2), \cdots, \rho_k((q + 1) \cdot 2^{p-k+1})]$, $q =$

$0, 1, \cdots, 2^{k-1}-1$ corresponds to an interchange of row $Row_{q\cdot 2^{p-k+1}+1}(H)$ with row $Row_{(q+1)\cdot 2^{p-k+1}}(H)$. If the condition (4.63) is satisfied so that the output variables $\mathcal{A}_{k-1,a} = \{Y_{j_1}, Y_{j_2}, \cdots, Y_{j_{k-1}}\}$ can be selected to determine reducible state variables, then there is

$$
\begin{aligned}
\sum_{i=q\cdot 2^{p-k+1}+1}^{(q+1)\cdot 2^{p-k+1}} v(\theta_k(i)) = & \sum_{i=q\cdot 2^{p-k+1}+1}^{(q+1)\cdot 2^{p-k+1}} v(\theta_{k-1}(\rho_k(i))) \\
= & \sum_{i=q\cdot 2^{p-k+1}+1}^{q\cdot 2^{p-k+1}+2^{p-k}} v(\theta_{k-1}(\rho_k(i))) + \sum_{i=q\cdot 2^{p-k+1}+2^{p-k}+1}^{(q+1)\cdot 2^{p-k+1}} v(\theta_{k-1}(\rho_k(i))) \\
= & 2^{n-k+1}.
\end{aligned} \tag{4.66}
$$

Hence, if $\sum_{i=q\cdot 2^{p-k+1}+1}^{q\cdot 2^{p-k+1}+2^{p-k}} v(\theta_{k-1}(\rho_k(i))) = 2^{n-k}$ holds, then it is clear that

$$
\sum_{i=q\cdot 2^{p-k+1}+2^{p-k}+1}^{(q+1)\cdot 2^{p-k+1}} v(\theta_{k-1}(\rho_k(i))) = 2^{n-k}.
$$

Therefore, if the condition (4.63) is satisfied, then instead of the condition (4.65) one should check

$$
\begin{cases}
\sum_{i=1}^{2^{p-k}} v(\theta_{k-1}(\rho_k(i))) = 2^{n-k}, \\
\sum_{i=2^{p-k+1}+1}^{2^{p-k+1}+2^{p-k}} v(\theta_{k-1}(\rho_k(i))) = 2^{n-k}, \\
\vdots \\
\sum_{i=2^p-2^{p-k+1}+1}^{2^p-2^{p-k}} v(\theta_{k-1}(\rho_k(i))) = 2^{n-k}.
\end{cases} \tag{4.67}
$$

Note that (4.67) has a smaller number of equalities. There are 2^{k-1} equalities in (4.67) while there are 2^k equalities in (4.65). Hence, finding the output Y_{j_k} based on (4.67) requires less computational effort. In case of fulfillment of (4.67), let $\Xi_k = \Xi_k \cup \{\mathcal{A}_{k-1,a} \cup \mathcal{A}_{k-1,b}\}$.

The procedure should be repeated until no new output variable can be found additionally or the maximum number of reducible state variables τ is reached according to Theorem 4.18. For convenience, the result above is summarized in Algorithm 2.

Algorithm 2: Given the BCN described by (2.34)–(2.35). Find the maximum set of output variables $\{Y_{j_1}, Y_{j_2}, \cdots, Y_{j_\tau}\}$ to determine reducible state variables.

1. Get the number of non-zero entries in each row of H, denoted by $v(1), v(2), \cdots, v(2^p)$.
2. Calculate τ_{max} according to (4.57). If $\tau_{max} = 0$, stop. Otherwise, proceed to Step 3.
3. Let $j_1 = 1$. Initialize $\widetilde{W}_0 = I_{2^p}$ and $\Xi_1 = \emptyset$.
4. Calculate $W_{[2,2^{j_1-1}]} \otimes I_{2^{p-j_1}}$, obtain $\rho_1(1), \rho_1(2), \cdots, \rho_1(2^p)$ according to the position of the non-zero entry in each column of the matrix $W_{[2,2^{j_1-1}]} \otimes I_{2^{p-j_1}}$.
5. Calculate $\widetilde{W}_1$ and obtain $\theta_1(1), \theta_1(2), \cdots, \theta_1(2^p)$ according to (4.50) and (4.51).
6. If (4.61) is satisfied, then let $\Xi_1 = \Xi_1 \cup \{Y_{j_1}\}$.
7. If $j_1 < p$, then let $j_1 = j_1 + 1$ and return to Step 2.
8. Let $k = 2$.
9. Initialize $\Xi_k = \emptyset$.
10. For all sets $\mathcal{A}_{k-1,a}, \mathcal{A}_{k-1,b} \in \Xi_{k-1}$, if (4.64) is satisfied and $|\mathcal{A}_{k-1,a} \cup \mathcal{A}_{k-1,b}| \notin \Xi_k$, calculate $I_{2^{k-1}} \otimes W_{[2,2^{j_k-k}]} \otimes I_{2^{p-j_k}}$ corresponding to $Y_{j_k} \in \mathcal{A}_{k-1,a} \backslash \mathcal{A}_{k-1,b}$ and get $\rho_k(1), \rho_k(2), \cdots, \rho_k(2^p)$.
11. Check (4.67) with $\theta_{k-1}(1), \theta_{k-1}(2), \cdots, \theta_{k-1}(2^p)$ obtained according to the candidate set $\mathcal{A}_{k-1,a}$ and $\rho_k(1), \rho_k(2), \cdots, \rho_k(2^p)$ obtained in Step 10.
12. If (4.67) is satisfied, then let $\Xi_k = \Xi_k \cup \{\mathcal{A}_{k-1,a} \cup \mathcal{A}_{k-1,b}\}$.
13. If no sets $\mathcal{A}_{k-1,a}$ and $\mathcal{A}_{k-1,b}$ in Step 10 satisfy (4.64) or $k > \tau_{max}$, stop and return the set Ξ_{k-1}. Otherwise replace k by $k+1$ and repeat the procedure from Step 10 to Step 13.

Remark 4.19 The computational complexity of an algorithm is the amount of resources required for running it in the worst case. Instead of Algorithm 2, a greedy algorithm may be applied to select the maximum number of output variables which makes the locally optimal choice at each step, i.e. chooses one output variable additionally at one time. However, a global optimal solution cannot generally be guaranteed by a greedy algorithm.

Once a set of output variables $\{Y_{j_1}, Y_{j_2}, \cdots, Y_{j_\tau}\}$ is found, the σ-swap matrix W_σ is set to

$$W_\sigma = \widetilde{W}_\tau^{\mathrm{T}} = \left(\delta_{2^p}[\theta_\tau(1)\ \theta_\tau(2)\ \cdots\ \theta_\tau(2^p)]\right)^{\mathrm{T}}. \tag{4.68}$$

As $\widetilde{W}_\tau$ is calculated recursively by applying (4.50), it is clear that for any $i = 1, 2, \cdots, 2^p$

$$\begin{aligned}\theta_\tau(i) &= \theta_{\tau-1}(\rho_\tau(i)) \\ &= \theta_{\tau-2}(\rho_{\tau-1}(\rho_\tau(i))) \\ &\vdots \\ &= \rho_1(\rho_2(\cdots(\rho_\tau(i))\cdots))\end{aligned} \tag{4.69}$$

holds where $\rho_k,\ k = 1, 2, \cdots, \tau$ are obtained by calculating $I_{2^{k-1}} \otimes W_{[2,2^{j_k-k}]} \otimes I_{2^{p-j_k}}$ based on the indices $j_1, j_2, \cdots, j_\tau$. One possible solution is to initialize the matrix W_σ as $\mathbf{0}_{2^p} \otimes \mathbf{0}_{2^p}^{\mathrm{T}}$. Then let the $(i, \theta_\tau(i))$-th entry of the matrix W_σ be set to 1 for $i = 1, 2, \cdots, 2^p$.

Determine the state transformation matrix T_G
After that, the permutation matrix T_G will be constructed. As a permutation matrix has exactly one entry of 1 in each row and column, the matrix T_G can be represented as

$$T_G^{\mathrm{T}} = \delta_{2^n}[\rho_x(1)\ \rho_x(2)\ \cdots\ \rho_x(2^n)] \tag{4.70}$$

where $\rho_x(i)$ is the position of a nonzero entry in the i-th column of T_G^{T}. From (4.70) it is clear that once $\rho_x(i), i = 1, 2, \cdots, 2^n$ is specified, the matrix T_G is known.

Recall that each block $\bar{H}_i, i = 1, 2, \cdots, 2^\tau$ is a logical matrix. By looking at (4.30) it can be recognized that the non-zero entries in the i-th block, i.e., $Row_j(W_\sigma H T_G^{\mathrm{T}}), j = i \cdot 2^{p-r} + 1, i \cdot 2^{p-r} + 2, \cdots, (i+1) \cdot 2^{p-r}$, shall lie in the columns $Col_k(W_\sigma H T_G^{\mathrm{T}}), k = i \cdot 2^{n-r} + 1, i \cdot 2^{n-r} + 2, \cdots, (i+1) \cdot 2^{n-r}$. Assume that in $Row_i(W_\sigma H)$ there are β_i non-zero entries whose positions in the row are denoted as $d_i^1, d_i^2, \cdots, d_i^{\beta_i}$. After right multiplication with T_G^{T}, one gets that

$$W_\sigma H T_G^{\mathrm{T}} \delta_{2^n}^k = W_\sigma H Col_k(T_G^{\mathrm{T}}) = W_\sigma H \delta_{2^n}^{\rho_x(k)} = Col_{\rho_x(k)}(W_\sigma H). \tag{4.71}$$

If $\rho_x(1), \rho_x(2), \cdots, \rho_x(2^n)$ are set to

$$\begin{array}{llll}\rho_x(1) = d_1^1, & \rho_x(\beta_1 + 1) = d_2^1, & \cdots, & \rho_x(\sum_{i=1}^{2P-1}\beta_i + 1) = d_{2^p}^1, \\ \rho_x(2) = d_1^2, & \rho_x(\beta_1 + 2) = d_2^2, & \cdots, & \rho_x(\sum_{i=1}^{2P-1}\beta_i + 2) = d_{2^p}^2, \\ \vdots & \vdots & \ddots & \vdots \\ \rho_x(\beta_1) = d_1^{\beta_1}, & \rho_x(\beta_1 + \beta_2) = d_2^{\beta_2}, & \cdots, & \rho_x(\sum_{i=1}^{2^p}\beta_i) = d_{2^p}^{\beta_{2^p}}\end{array} \tag{4.72}$$

and after the matrix $\tilde{H} = W_\sigma H T_G^{\mathrm{T}}$ is partitioned as (4.32), then it can be verified that only the blocks $\tilde{H}_{ii}, i = 1, 2, \cdots, 2^\tau$ are non-zero matrices. That means the matrix $\tilde{H}$ has the structure shown in (4.30).

For any i after permuting the numbers $d_i^1, d_i^2, \cdots, d_i^{\beta_i}$, the resulting T_G together with W_σ can still transform the matrix H into the form of (4.30). Hence, the solution for the matrix T_G is not unique. As a matter of fact, there is a total of $\beta_1! \cdot \beta_2! \cdot \cdots \cdot \beta_{2^p}!$ equivalent solutions.

Example 4.20 In order to illustrate the main results in this section, consider the following BCN

$$\begin{cases} X_1(t+1) = \neg X_3(t) \wedge (X_1(t) \vee U(t)), \\ X_2(t+1) = \neg X_4(t) \wedge (X_1(t) \vee X_3(t)), \\ X_3(t+1) = X_2(t), \\ X_4(t+1) = \neg X_1(t) \wedge (X_2(t) \vee X_3(t)) \end{cases} \tag{4.73}$$

which is a Boolean network of the $p53$ pathway given in Layek et al. (2011).

To illustrate the procedure in Algorithm 2, assume that the output equations are given as

$$\begin{cases} Y_1(t) = X_4 \vee \neg X_3 \wedge (\neg X_1 \vee X_2), \\ Y_2(t) = X_4 \vee X_2 \wedge \neg X_3 \vee \neg X_1 \wedge \neg X_2 \vee \\ \qquad X_1 \wedge \neg X_2 \wedge (X_3 \vee \neg X_4) \wedge (\neg X_3 \vee X_4), \\ Y_3(t) = \neg X_1 \wedge (X_2 \vee \neg X_4). \end{cases} \tag{4.74}$$

Using the STP and applying the vector form of Boolean variables, the BCN (4.73)–(4.74) can be converted into the algebraic form (2.34)–(2.35) with $x(t) \in \Delta_{16}$, $u(t) \in \Delta_2$, $y(t) \in \Delta_8$ and

$$\begin{aligned} L_{eq} &= \delta_{16}[14\ 14\ 10\ 10\ 6\ 6\ 2\ 2\ 16\ 16\ 12\ 12\ 8\ 8\ 4\ 4 \\ &\qquad 13\ 13\ 9\ 9\ 5\ 13\ 5\ 13\ 15\ 15\ 11\ 11\ 8\ 16\ 8\ 16], \\ H &= \delta_8[8\ 7\ 8\ 1\ 6\ 7\ 4\ 5\ 1\ 7\ 1\ 1\ 6\ 2\ 2\ 2]. \end{aligned}$$

At first, the output variables to determine the reducible state variables will be found. The number of nonzero entries in each row of the matrix H are, respectively, $v(1) = 4$, $v(2) = 3$, $v(3) = 0$, $v(4) = 1$, $v(5) = 1$, $v(6) = 2$, $v(7) = 3$, $v(8) = 2$. In the first step, the candidate output variable Y_{j_1} will be determined. Let the output variable Y_1 be considered as a candidate output variable. The matrix $\widetilde{W}_1$ corresponding to $j_1 = 1$ is calculated by (4.50) which results in

$$\begin{aligned}\widetilde{W}_1 =& \delta_8[\theta_1(1)\ \theta_1(2)\ \theta_1(3)\ \theta_1(4)\ \theta_1(5)\ \theta_1(6)\ \theta_1(7)\ \theta_1(8)] \\ =& \delta_8[1\ 2\ 3\ 4\ 5\ 6\ 7\ 8]. \end{aligned} \tag{4.75}$$

In order to check the condition (4.61), the total number of the non-zero entries in rows 1, 2, 3, 4 in the matrix H should be calculated. That is $\upsilon(\theta_1(1)) + \upsilon(\theta_1(2)) + \upsilon(\theta_1(3)) + \upsilon(\theta_1(4)) = 4 + 3 + 0 + 1 = 8$ which is equal to $16/2$. Therefore, the output variable Y_1 is a candidate. In a similar way, for the output variables Y_2 and Y_3 the rows 1, 2, 5, 6 and 1, 3, 5, 7 in the matrix H are, respectively, taken into account. The number of non-zero entries in these rows are, respectively, $4 + 3 + 1 + 2 = 10$ and $4 + 0 + 1 + 3 = 8$. As $10 \neq 16/2$, the output variable Y_2 cannot be selected as a candidate. So the candidate output variables are Y_1 and Y_3 and the set Ξ_1 is $\Xi_1 = \{\mathcal{A}_{1,1}, \mathcal{A}_{1,2}\} = \{\{Y_1\}, \{Y_3\}\}$. After that, as the set $|\{Y_1\} \cap \{Y_3\}| = 1 - 1 = 0$ satisfies (4.64), the set of output variables $\{Y_1, Y_3\}$ will be checked. Calculate the matrix $I_2 \otimes W_{[2,2]}$ and one has

$$\begin{aligned} I_2 \otimes W_{[2,2]} &= \delta_8[\rho_2(1)\ \rho_2(2)\ \rho_2(3)\ \rho_2(4)\ \rho_2(5)\ \rho_2(6)\ \rho_2(7)\ \rho_2(8)] \\ &= \delta_8[1\ 3\ 2\ 4\ 5\ 7\ 6\ 8]. \end{aligned} \tag{4.76}$$

Then, the corresponding matrix $\widetilde{W}_2$ can be calculated by (4.50). A simple computation shows that

$$\begin{aligned}\widetilde{W}_2 =& \delta_8[\theta_2(1)\ \theta_2(2)\ \theta_2(3)\ \theta_2(4)\ \theta_2(5)\ \theta_2(6)\ \theta_2(7)\ \theta_2(8)] \\ =& \delta_8[1\ 3\ 2\ 4\ 5\ 7\ 6\ 8]. \end{aligned} \tag{4.77}$$

According to (4.67), one should check the total number of non-zero entries in rows 1, 3 and 5, 7 in the matrix H. The number of non-zero entries is $\upsilon(\theta_2(1)) + \upsilon(\theta_2(2)) = 4 + 0 = 4$ and $\upsilon(\theta_2(5)) + \upsilon(\theta_2(6)) = 1 + 3 = 4$ which are equal to $16/4 = 4$. According to the condition (4.67), the set of output variables $\{Y_1, Y_3\}$ can be used to determine reducible state variables and the set $\Xi_2 = \{\{Y_1, Y_3\}\}$. As there is only one candidate set in Ξ_2, the procedure can be stopped and $\tau = 2$.

In the next step, the transformation matrices T_G and W_σ will be found. As Y_1 and Y_3 are the output variables to determine two reducible state variables, according to (4.68) the transformation matrix for output W_σ is set to $W_\sigma = \widetilde{W}_2^{\mathrm{T}} = \delta_8[1\ 3\ 2\ 4\ 5\ 7\ 6\ 8]$ and the matrix $W_\sigma H$ is

$$W_\sigma H = \delta_8[8\ 6\ 8\ 1\ 7\ 6\ 4\ 5\ 1\ 6\ 1\ 1\ 7\ 3\ 3\ 3]. \tag{4.78}$$

Let the permutation matrix T_G be expressed as $T_G^{\mathrm{T}} = \delta_{16}[\rho_x(1)\ \rho_x(2)\ \cdots\ \rho_x(16)]$. By looking at $Row_1(W_\sigma H)$, the non-zero entries are $[W_\sigma H]_{(1,4)} = 1, [W_\sigma H]_{(1,9)} = 1$, $[W_\sigma H]_{(1,11)} = 1$ and $[W_\sigma H]_{(1,12)} = 1$ which means that $d_1^1 = 4, d_1^2 = 9, d_1^3 = 11, d_1^4 = 12$. According to (4.72), $\rho_x(1) = d_1^1 = 4$, $\rho_x(2) = d_1^2 = 9$, $\rho_x(3) = d_1^3 = 11$ and $\rho_x(4) = d_1^4 = 12$ are obtained. As there are only zero entries in $Row_2(W_\sigma H)$, no permutation will be considered for $Row_2(W_\sigma H)$. Considering $Row_3(W_\sigma H)$, it is found out that $[W_\sigma H]_{(3,14)} = 1$, $[W_\sigma H]_{(3,15)} = 1$ and $[W_\sigma H]_{(3,16)} = 1$, i.e. $d_3^1 = 14, d_3^2 = 15, d_3^4 = 16$. Therefore, one has $\rho_x(5) = d_3^1 = 14$, $\rho_x(6) = d_3^2 = 15$ and $\rho_x(7) = d_3^3 = 16$. Following the same procedure for $Row_i(W_\sigma H), i = 4, 5, \cdots, 16$, there is

$$T_G^{\mathrm{T}} = \delta_{16}[4\ 9\ 11\ 12\ 14\ 15\ 16\ 7\ 8\ 2\ 6\ 10\ 5\ 13\ 1\ 3].$$

After that, the matrix $W_\sigma H T_G^{\mathrm{T}}$ is calculated. The result is

$$W_\sigma H T_G^{\mathrm{T}} = \left[\begin{array}{cccc|cccc|cccc|cccc} 1&1&1&1&0&0&0&0&0&0&0&0&0&0&0&0\\ 0&0&0&0&0&0&0&0&0&0&0&0&0&0&0&0\\ \hline 0&0&0&0&1&1&1&0&0&0&0&0&0&0&0&0\\ 0&0&0&0&0&0&0&1&0&0&0&0&0&0&0&0\\ \hline 0&0&0&0&0&0&0&0&1&0&0&0&0&0&0&0\\ 0&0&0&0&0&0&0&0&0&1&1&1&0&0&0&0\\ \hline 0&0&0&0&0&0&0&0&0&0&0&0&1&1&0&0\\ 0&0&0&0&0&0&0&0&0&0&0&0&0&0&1&1 \end{array}\right], \tag{4.79}$$

which has the structure of (4.30).

4.3.4 Observer Design and State Reconstruction

If the reducible state variables are found and the BCN (2.34)–(2.35) can be brought into the form of (4.24)–(4.26) by state and output transformation, then a reduced-order observer design can be carried out as introduced in this section. At first, an approach to design a reduced-order observer for the BCN will be proposed. After that, the performance of the reduced-order observer will be studied.

Assume that the output variables $Y_{j_1}, Y_{j_2}, \cdots, Y_{j_\tau}$ have been found to determine reducible state variables based on Algorithm 2 proposed in Subsection 4.3.3. According to (4.21), it is clear that

$$\tilde{y}_1(t) = \ltimes_{i=1}^{\tau} y_{j_i}(t),\ \forall t = 0, 1, \cdots. \tag{4.80}$$

For the BCN described by (4.24)–(4.26) a reduced-order observer is constructed as

$$\hat{\tilde{z}}_2(t) = \mathbf{1}_{2^\tau}^{\mathrm{T}} \Phi_{2^n}^{\mathrm{T}} (I_{2^n} \otimes (T_G H^{\mathrm{T}}))\tilde{L}\tilde{y}_1(t-1)\hat{\tilde{z}}_2(t-1)u(t-1)y(t),\ t = 1, 2, \cdots \tag{4.81}$$

where $\tilde{y}_1(t-1)$ is obtained by (4.80) and

$$\hat{\tilde{z}}_2(0) = \mathbf{1}_{2^\tau}^{\mathrm{T}} T_G H^{\mathrm{T}} y(0). \tag{4.82}$$

Based on the state estimate $\hat{\tilde{z}}_2(t)$ provided by the reduced-order observer (4.81), the state of the BCN (2.32)–(2.33) can be reconstructed in two steps:

1. Reconstruction of the state $\hat{z}(t)$: Because of (4.25), $z(t) = \tilde{z}_1(t)\tilde{z}_2(t) = \tilde{y}_1(t)\tilde{z}_2(t)$. Hence, after having obtained the output $\tilde{y}_1(t)$ by (4.80) and the estimate $\hat{\tilde{z}}_2(t)$ by (4.81), the state $\hat{z}(t)$ can be reconstructed as

$$\hat{z}(t) = \tilde{y}_1(t)\hat{\tilde{z}}_2(t). \tag{4.83}$$

2. Transformation from the state $\hat{z}(t)$ to the state $\hat{x}(t)$: Recalling the transformation matrix T_G given in (6.51) and as the matrix T_G is a permutation matrix (i.e. $T_G^{-1} = T_G^{\mathrm{T}}$), the state estimate $\hat{x}(t)$ can be obtained by

$$\hat{x}(t) = T_G^{\mathrm{T}} \hat{z}(t). \tag{4.84}$$

Remark 4.21 For the state estimation of a large-scale BCN, the calculation of the system matrix $S = \mathbf{1}_{2^\tau}^{\mathrm{T}} \Phi_{2^n}^{\mathrm{T}} (I_{2^n} \otimes (T_G H^{\mathrm{T}}))\tilde{L}$ of the reduced-order observer (4.81) may be infeasible. To solve this problem, recall (4.70) and let $\delta_{2^n}^{\pi_x(i)} = \delta_{2^\tau}^{i_1} \delta_{2^{n-\tau}}^{i_2}$ and $w = (\pi_x(k) - 1)2^{m+p} + (l-1)2^p + j$. For the sake of compactness, in this remark the notation $\delta_{2^n}^i$ represents the corresponding values of Boolean variables. Let $f(\delta_{2^n}^k, \delta_{2^m}^l) = [f_1(\delta_{2^n}^k, \delta_{2^m}^l)\ f_2(\delta_{2^n}^k, \delta_{2^m}^l)\ \cdots\ f_n(\delta_{2^n}^k, \delta_{2^m}^l)]^{\mathrm{T}}$ and $h(\delta_{2^n}^i) = [h_1(\delta_{2^n}^i)\ h_2(\delta_{2^n}^i)\ \cdots\ h_p(\delta_{2^n}^i)]^{\mathrm{T}}$. An alternative way that requires lower computational effort to construct the matrix S is

$$S_{(i_2,w)} = \begin{cases} 1, & \text{if } \delta_{2^n}^i = f(\delta_{2^n}^k, \delta_{2^m}^l) \text{ and } \delta_{2^p}^j = h(\delta_{2^n}^i), \\ 0, & \text{otherwise} \end{cases}$$

where $\delta_{2^n}^i = f(\delta_{2^n}^k, \delta_{2^m}^l)$ implies that the input $\delta_{2^m}^l$ steers the state $\delta_{2^n}^k$ to the state $\delta_{2^n}^i$. Similarly, $\delta_{2^p}^j = h(\delta_{2^n}^i)$ shows that the state $\delta_{2^n}^i$ can generate the output $\delta_{2^p}^j$.

4.3.5 Observer Performance

The main goal of this subsection is to show that the state estimate of the BCN (2.34)–(2.35), reconstructed based on the reduced-order observer (4.81)–(4.84) is equivalent to the state estimate provided by the Luenberger-like observer introduced in Theorem 4.3.

Reconstructibility is the property of a system that allows the current state to be determined uniquely with the knowledge of the input and output trajectory. Recall that the BCN (2.34)–(2.35) is reconstructible in the sense of Definition 3.7. Hence, the state estimate $\hat{x}_s(t)$ delivered by the Luenberger-like observer (6.70) converges to the real state $x(t)$ no later than the minimal reconstructibility index r_{min}.

After having obtained the state estimate $\hat{x}(t)$ provided by the reduced-order observer (4.81)–(4.84), the following conclusion can be obtained.

Theorem 4.22 *The state estimate $\hat{x}(t)$ provided by the reduced-order observers* (4.81)–(4.84) *is always equal to the state estimate $\hat{x}_s(t)$ delivered by the Luenberger-like observer* (4.4). *If the BCN (2.34)–(2.35) is reconstructible, then the state estimate $\hat{x}(t)$ converges to the real state (i.e. $\hat{x}(t) = x(t)$) no later than time $t = r_{min}$.*

Proof. As permutation matrices T_G and W_σ are orthogonal matrices, one has

$$T_G^{\mathrm{T}} T_G = I_{2^n} \text{ and } W_\sigma^{\mathrm{T}} W_\sigma = I_{2^p}. \tag{4.85}$$

According to the output transformation (4.21) and considering (4.85), there is

$$\tilde{H}^{\mathrm{T}} \tilde{y}(t) = T_G H^{\mathrm{T}} W_\sigma^{\mathrm{T}} W_\sigma y(t) = T_G H^{\mathrm{T}} y(t), \forall t = 0, 1, \cdots. \tag{4.86}$$

By recalling (4.25), i.e. $\tilde{y}_1(t) = \tilde{z}_1(t), \forall t = 0, 1, \cdots$, it follows that $(I_{2^\tau} \otimes \mathbf{1}_{2^{n-\tau}}^{\mathrm{T}}) \tilde{H}^{\mathrm{T}} \tilde{y}(t) = \tilde{y}_1(t)$. After applying state and output transformation, the BCN (2.34)–(2.35) takes the same form as (4.24)–(4.26). Based on this, it can be directly understood that for $t = 0, 1, \cdots$ it holds

$$\begin{aligned}\tilde{H}^{\mathrm{T}}\tilde{y}(t) &= \left((I_{2^\tau} \otimes \mathbf{1}_{2^{n-\tau}}^{\mathrm{T}})\tilde{H}^{\mathrm{T}}\tilde{y}(t)\right)\left(\mathbf{1}_{2^\tau}^{\mathrm{T}}\tilde{H}^{\mathrm{T}}\tilde{y}(t)\right) \\ &= \tilde{y}_1(t)\mathbf{1}_{2^\tau}^{\mathrm{T}}\tilde{H}^{\mathrm{T}}\tilde{y}(t).\end{aligned} \tag{4.87}$$

Now the theorem will be proven by induction. Let start with time $t = 0$. Based on (4.82)–(4.83) and applying (4.85)–(4.86), (4.84) can be rewritten as

$$\begin{aligned}\hat{x}(0) &= T_G^{\mathrm{T}}\tilde{y}_1(0)\hat{\tilde{z}}_2(0) \\ &= T_G^{\mathrm{T}}\tilde{y}_1(0)\mathbf{1}_{2^\tau}^{\mathrm{T}}\tilde{H}^{\mathrm{T}}\tilde{y}(0) = T_G^{\mathrm{T}}\tilde{H}^{\mathrm{T}}\tilde{y}(0) \\ &= T_G^{\mathrm{T}}T_G H^{\mathrm{T}}W_\sigma^{\mathrm{T}}W_\sigma y(0) \\ &= H^{\mathrm{T}}y(0) \\ &= \hat{x}_s(0)\end{aligned} \tag{4.88}$$

which shows that the state estimate $\hat{x}(0)$ provided by the reduced-order observers is equal to the state estimate $\hat{x}_s(0)$ provided by the Luenberger-like observer (6.70).

Next, assume that $\hat{x}(t) = \hat{x}_s(t)$ holds for $t = k$. According to (4.25) and (4.86) and based on (4.81), (4.83) and (4.84), there is

$$\begin{aligned}\hat{x}(k+1) &= T_G^{\mathrm{T}}\tilde{y}_1(k+1)\hat{\tilde{z}}_2(k+1) \\ &= T_G^{\mathrm{T}}\tilde{y}_1(k+1)\mathbf{1}_{2^\tau}^{\mathrm{T}}\Phi_{2^n}^{\mathrm{T}}(I_{2^n} \otimes (T_G H^{\mathrm{T}}))\tilde{L}\tilde{y}_1(k)\hat{\tilde{z}}_2(k)u(k)y(k+1) \\ &= T_G^{\mathrm{T}}\tilde{y}_1(k+1)\mathbf{1}_{2^\tau}^{\mathrm{T}}\Phi_{2^n}^{\mathrm{T}}\tilde{L}\hat{z}(k)u(k)T_G H^{\mathrm{T}}y(k+1) \\ &= T_G^{\mathrm{T}}\tilde{y}_1(k+1)\mathbf{1}_{2^\tau}^{\mathrm{T}}\Phi_{2^n}^{\mathrm{T}}\tilde{L}\hat{z}(k)u(k)T_G T_G^{\mathrm{T}}T_G H^{\mathrm{T}}W_\sigma^{\mathrm{T}}W_\sigma y(k+1) \\ &= T_G^{\mathrm{T}}\tilde{y}_1(k+1)\mathbf{1}_{2^\tau}^{\mathrm{T}}\Phi_{2^n}^{\mathrm{T}}\tilde{L}\hat{z}(k)u(k)\tilde{H}^{\mathrm{T}}\tilde{y}(k+1).\end{aligned} \tag{4.89}$$

As (4.25) is always satisfied, one can conclude that there is an index i such that $[\hat{z}(k)]_i \neq 0$ and $(I_{2^\tau} \otimes \mathbf{1}_{2^{n-\tau}}^{\mathrm{T}})\tilde{L}\delta_{2^n}^i u(k) = \tilde{y}_1(k+1)$. According to Lemma 2.10 and (4.87), it follows that

$$\Phi_{2^n}^{\mathrm{T}}\tilde{L}\hat{z}(k)u(k)\tilde{H}^{\mathrm{T}}\tilde{y}(k+1) = \tilde{y}_1(k+1)\mathbf{1}_{2^\tau}^{\mathrm{T}}\Phi_{2^n}^{\mathrm{T}}\tilde{L}\hat{z}(k)u(k)\tilde{H}^{\mathrm{T}}\tilde{y}(k+1). \tag{4.90}$$

By applying (4.90) and recalling $\tilde{L} = T_G L T_G^{\mathrm{T}}$, (4.89) can be simplified as

$$
\begin{aligned}
\hat{x}(k+1) &= T_G^{\mathrm{T}}\tilde{y}_1(k+1)\mathbf{1}_{2^\tau}^{\mathrm{T}}\tilde{y}_1(k+1)\mathbf{1}_{2^\tau}^{\mathrm{T}}\Phi_{2^n}^{\mathrm{T}}\tilde{L}\hat{z}(k)u(k)\tilde{H}^{\mathrm{T}}\tilde{y}(k+1)\\
&= T_G^{\mathrm{T}}\tilde{y}_1(k+1)\mathbf{1}_{2^\tau}^{\mathrm{T}}\Phi_{2^n}^{\mathrm{T}}\tilde{L}\hat{z}(k)u(k)\tilde{H}^{\mathrm{T}}\tilde{y}(k+1)\\
&= T_G^{\mathrm{T}}\Phi_{2^n}^{\mathrm{T}}\tilde{L}\hat{z}(k)u(k)\tilde{H}^{\mathrm{T}}\tilde{y}(k+1)\\
&= T_G^{\mathrm{T}}\Phi_{2^n}^{\mathrm{T}}\left(T_G L T_G^{\mathrm{T}}\hat{z}(k)u(k)\right)\left(T_G H^{\mathrm{T}} W_\sigma^{\mathrm{T}}\tilde{y}(k+1)\right)\\
&= T_G^{\mathrm{T}}T_G\Phi_{2^n}^{\mathrm{T}}L T_G^{\mathrm{T}}\hat{z}(k)u(k)H^{\mathrm{T}}y(k+1).
\end{aligned}
\tag{4.91}
$$

If $\hat{x}(k) = \hat{x}_s(k)$, then $\hat{z}(k) = T_G\hat{x}(k) = T_G\hat{x}_s(k)$. Replacing $\hat{z}(k)$ in (4.91) by $T_G\hat{x}_s(k)$ and applying (4.85), (4.91) can be rewritten as

$$
\begin{aligned}
\hat{x}(k+1) &= T_G^{\mathrm{T}}T_G\Phi_{2^n}^{\mathrm{T}}L T_G^{\mathrm{T}}T_G\hat{x}_s(k)u(k)H^{\mathrm{T}}y(k+1)\\
&= \Phi_{2^n}^{\mathrm{T}}L\hat{x}_s(k)u(k)H^{\mathrm{T}}y(k+1)\\
&= \hat{x}_s(k+1)
\end{aligned}
$$

which shows that $\hat{x}(t) = \hat{x}_s(t)$ holds also for $t = k+1$.

By the principle of induction, it can be seen that $\hat{x}(t) = \hat{x}_s(t)$ holds for any t. That means the state estimate $\hat{x}(t)$ provided by the reduced-order observer is equal to the state estimate $\hat{x}_s(t)$ provided by the Luenberger-like observer.

It is worth noting that the state estimate $\hat{x}_s(t)$ provided by the Luenberger-like observer (4.4) converges to the real state no later than the minimal reconstructibility index r_{min}. As $\hat{x}(t) = \hat{x}_s(t)$ holds for any t, it is clear that $\hat{x}(t)$ provided by the reduced-order observer (4.81)–(4.84) has the same property and will also converge to the real state not later than time $t = r_{min}$. □

Remark 4.23 The reduced-order observer (4.81)–(4.84) requires much lower online computational effort and memory consumption than the Luenberger-like observer (4.4). The computational complexity of the reduced-order observer (4.81)–(4.84) is $\mathcal{O}(2^{2n+m+p-\tau})$ where τ is the number of reducible state variables. In comparison, the computational burden needed by the Luenberger-like observer (4.4) is $\mathcal{O}(2^{2n+m+p})$.

Remark 4.24 The reduced order observer (4.82)–(4.84) is not only suitable for online implementation due to significantly reduced lower computational effort, but also applicable for the observer-based fault detection of BCNs with lower computational effort, for instance to test whether a meaningful fault occurred (Fornasini and Valcher, 2015a). For this purpose, the reduced-order observer is initialized as (4.82). Let the reduced-order observer repeat the recursive procedure to estimate $\tilde{z}_2(t)$. If at some time t the state estimate $\hat{\tilde{z}}_2(t)$ provided by the reduced-order observer becomes a zero vector (i.e. $\hat{\tilde{z}}_2(t) = \mathbf{0}_{2^{n-\tau}}$), it is clear that $\hat{x}(t) = T_G \tilde{y}_1(t)\hat{\tilde{z}}_2(t) = \mathbf{0}_{2^{n-\tau}}$. This means that no state can be compatible with the input and output trajectory. In other words, the input and output trajectory is inconsistent with normal system behavior. Hence, it can be concluded that a fault has occurred. However, if $\hat{\tilde{z}}_2(t)$ is not a zero vector, then one can conclude that so far no meaning fault has affected the BCN's behavior.

Example 4.25 In order to show the performance of the reduced-order observer, a Boolean model for oxidative stress response pathways is considered which is derived from Sridharan et al. (2012):

$$\begin{cases} X_1(t+1) = U(t) \wedge \neg X_6(t), \\ X_2(t+1) = \neg X_1(t), \\ X_3(t+1) = \neg X_1(t) \wedge (X_5(t) \vee X_3(t)), \\ X_4(t+1) = X_1(t) \wedge \neg X_6(t), \\ X_5(t+1) = X_4(t) \vee \neg X_3(t), \\ X_6(t+1) = X_5(t) \wedge (\neg X_6(t) \vee \neg X_2(t)). \end{cases} \tag{4.92}$$

Assume that output equations are

$$Y_1(t) = X_2(t), \quad Y_2(t) = X_3(t), \quad Y_3(t) = X_5(t). \tag{4.93}$$

Letting $x(t) \in \Delta_{64}$, $u(t) \in \Delta_2$, $y(t) \in \Delta_8$ and using STP, the BCN (4.92)–(4.93) is expressed in the matrix form (2.32)–(2.33) with $L \in \mathcal{L}_{64\times128}$, $H \in \mathcal{L}_{8\times64}$.

In the same way as in Example 4.20, three reducible state variables are found and the permutation matrices W_σ and T_G are determined. The reconstructibility analysis using the approach proposed in Section 3.2 shows that the BCN (4.92)–(4.93) is reconstructible and the minimal reconstructibility index is $r_{min} = 4$.

Table 4.2 Input and output trajectory in vector form

t	0	1	2	3	4	5	6	7	8	9	10
u	δ_2^1	δ_2^1	δ_2^1	δ_2^1	δ_2^2	δ_2^2	δ_2^1	δ_2^1	δ_2^2	δ_2^1	
y	δ_8^8	δ_8^3	δ_8^1	δ_8^8	δ_8^3	δ_8^7	δ_8^1	δ_8^2	δ_8^2	δ_8^8	δ_8^3

In the next step, the reduced-order observer is constructed as shown in (4.81)–(4.84). Suppose that an input and output trajectory is given in Table 4.2. The state estimate of non-reducible state variables provided by the reduced-order observer is shown in Fig. 4.3. The performances of the reduced-order observer and the Luenberger-like observer are compared as shown in Fig. 4.4. The state estimate provided by the reduced-order observer converges to the correct state at time $t = 2$. Moreover, the state estimate $\hat{x}(t)$ based on the reduced-order observer is always equal to the state estimate $\hat{x}_s(t)$ of the Luenberger-like observer. As the reduced-order observer needs to evaluate only the part of state $\tilde{z}_2(t)$, the dimensions of system matrix in (4.81) is 8×1024. Besides, the state transformation matrix T_G in (4.84) has the dimensions 64×64. In contrast, the Luenberger-like observer (4.4) estimates the full state. For this purpose, a 64×1024 dimensional system matrix is applied which has a much higher dimension than the one used by the reduced-order observer. To show the running time of applying the reduced-order observer and the Luenberger-like observer, the procedure to evaluate state estimate in the interval $t \in [0, 10]$ is repeated 100 times using a CPU Intel Core i5-7500 @3.40GHz. As a result, the mean and variance of the CPU time needed for processing a state estimation by using the reduced-order observer are, respectively, $8.99 \times 10^{-4}s$ and $5.69 \times 10^{-8}s$ while for the Luenberger-like observer the mean and variance of the CPU time are $4.1 \times 10^{-3}s$ and $3.01 \times 10^{-7}s$. Hence, the reduced-order observer needs much less online calculation.

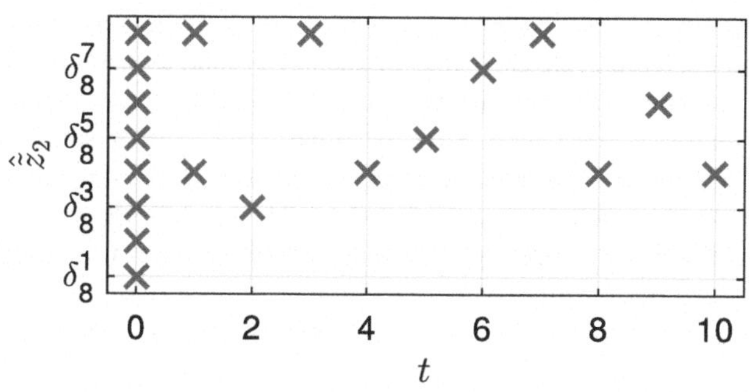

Figure 4.3 State estimate $\hat{\tilde{z}}_2(t)$ for state $\tilde{z}_2(t)$ provided by the reduced-order observer

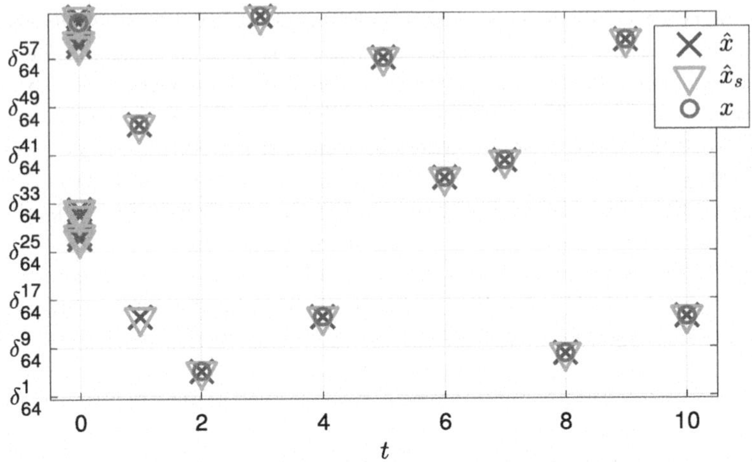

Figure 4.4 State estimate (x: real state, $\hat{x}$: state estimate obtained based on the reduced-order observer, $\hat{x}_s$: state estimate obtained by the Luenberger-like observer)

4.4 Distributed Observer Design

As large-scale systems consist of a large number of states, a centralized observer would possibly need a lot of computational effort. One possible way to cope with this is to partition the large-scale BCN into smaller BCNs and then apply the proposed observer design approach to each subsystem, i.e. use a distributed observer.

Assume that a large-scale BCN has been partitioned into α subnetworks. If a large-scale BCN is reconstructible according to Theorem 3.25, then a distributed observer can be designed for state estimation of large-scale BCNs. A distributed observer contains α local observers. The i-th local observer for the i-th subnetwork (3.52)–(3.53) can be constructed as

$$\begin{aligned}
&\hat{x}_{sub_i}(0) = H_{sub_i}^{\mathrm{T}} y_{sub_i}(0), \qquad (4.94)\\
&\hat{x}_{sub_i}(t) = \Phi_{2^{n_i}}^{\mathrm{T}}(I_{2^{n_i}} \otimes H_i^{\mathrm{T}}) L_{sub_i} \hat{x}_{sub_i}(t-1) u_{sub_i}(t-1) \hat{z}_i(t-1) y_{sub_i}(t),\ t = 1, 2, \cdots
\end{aligned}$$

where $y_{sub_i}(t)$ and $u_{sub_i}(t-1)$ are, respectively, the output and the known input of the i-th subnetwork, $\hat{x}_{sub_i}$ is the state estimation and $\hat{z}_i(t)$ is the information delivered from the local observers of the in-neighbors of the i-th subnetwork.

According to Theorem 3.25, if all conditions are satisfied, then the large-scale BCN can be regarded as directed acyclic graph. All α subnetworks are sorted into β levels denoted by the set Ξ_i, $i = 0, 1, \cdots, \beta-1$. The steps t_j, $j = 0, 1, \cdots, \beta-1$, are calculated according to (3.54).

Theorem 4.26 *If a large-scale BCN (2.13)–(2.14) is reconstructible according to Theorem 3.25, then the distributed observer containing α local observers described by* (4.94) *can provide the state estimate of the BCN without estimation error (i.e. $x_{sub_i}(t) = \hat{x}_{sub_i}(t), i = 1, 2, \cdots, \alpha$) within $t_{\beta-1}$ steps.*

Proof. Theorem 4.3 has pointed out that the observer (4.4) can always provide the correct state estimate of the BCN at a time greater than or equal to the minimal reconstructibility index. As t_0 denotes the largest number of minimal reconstructibility indices at level 0, the observers for state reconstruction of the subnetworks in set Ξ_0 can provide the correct state estimate within t_0 steps. After that, all inputs for the subnetworks in Ξ_1 at level 1 are known. According to Theorem 4.3, the observers for the subnetworks at level 1 give a state estimate equal to the correct state within t_1 steps. Repeating the procedure until level $\beta - 1$, all local observers can provide a state estimate equal to the correct state. Then, due to $\mathcal{X}_i \cap \mathcal{X}_j = \emptyset$ with $i \neq j$, the distributed observer for large-scale BCN provides the real state within $t_{\beta-1}$ steps.□

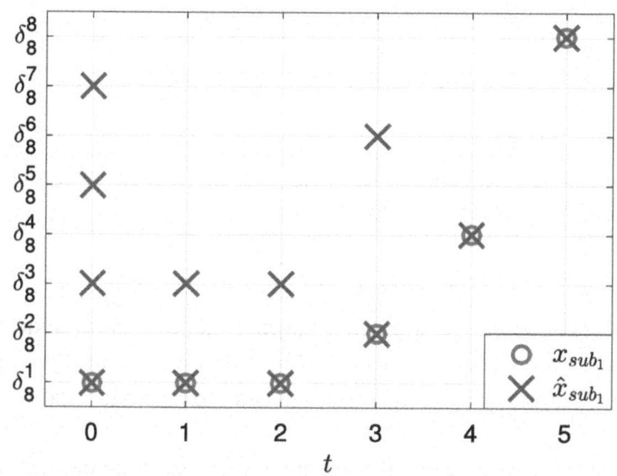

Figure 4.5 State estimate delivered by observer 1 for the subnetwork $Sub1$ ($x_{sub_1}(t)$: real state at time t, $\hat{x}_{sub_1}(t)$: state estimate at time t)

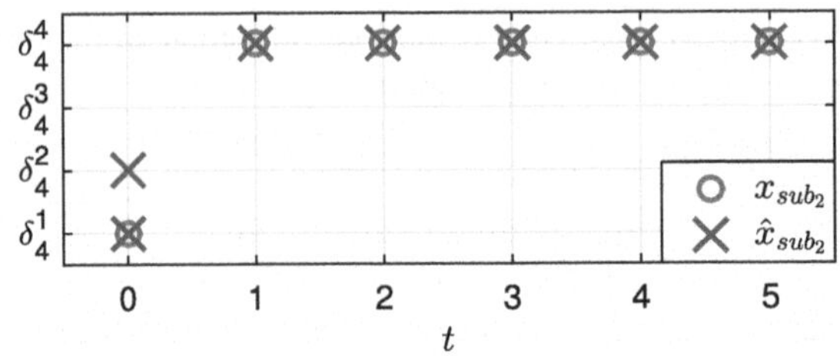

Figure 4.6 State estimate delivered by observer 2 for the subnetwork $Sub2$ ($x_{sub_2}(t)$: real state at time t, $\hat{x}_{sub_2}(t)$: state estimate at time t)

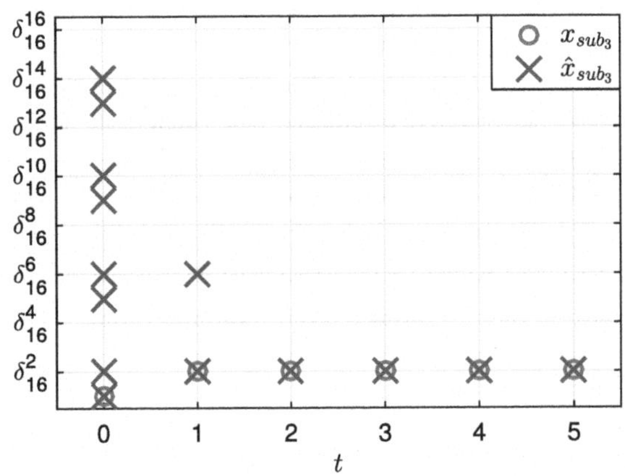

Figure 4.7 State estimate delivered by observer 3 for the subnetwork $Sub3$ ($x_{sub_3}(t)$: real state at time t, $\hat{x}_{sub_3}(t)$: state estimate at time t)

Example 4.27 Assume that an input and output trajectory is given in Table 4.3. For each subnetwork, an observer is created. The performance of observers is shown in Fig. 4.5–4.7. At time $t = 4$, all observers already provide the correct state of subnetworks.

Now consider the Luenberger-like observer (4.4) for the state estimation of the BCN (3.58). Assume that logical matrices $L \in \mathcal{L}_{512\times 2048}$ and $H \in \mathcal{L}_{8\times 512}$ are already known. $\hat{x}(0)$ is initialized as $H^{\mathrm{T}}y(0)$. According to (4.4), the Luenberger-like observer updates its estimated state, namely $\hat{x}(1)$, which reveals that δ_{512}^{50}, δ_{512}^{54},

Table 4.3 Input and output trajectory (logical value)

t	0	1	2	3	4	5
U_1	1	1	0	0	0	
U_2	0	1	1	0	0	
Y_1	1	1	1	0	0	0
Y_2	1	0	0	0	0	0
Y_3	1	1	1	1	1	1

δ_{512}^{178}, δ_{512}^{182} are possible states. Repeating the same procedure for $t = 2, 3, 4, 5$, the estimated states are $\hat{x}(2) = \delta_{512}^{50}$, $\hat{x}(3) = \delta_{512}^{114}$, $\hat{x}(4) = \delta_{512}^{242}$ and $\hat{x}(5) = \delta_{512}^{498}$. In comparison to the distributed observer, the Luenberger-like observer converges to the correct state earlier but needs much larger memory consumption.

Model-based output tracking control

5

This chapter will address the problem of finite horizon output tracking control of BCNs. To reach this goal, the first step is giving necessary and sufficient conditions for the trackability of a time-varying reference output trajectory of finite length. Then, an approach to determine a control sequence for a trackable reference output trajectory is proposed. If the reference output trajectory is not trackable, then two approaches are provided for the design of the optimal control sequence with the purpose of minimizing the tracking error, i.e. to formulate the control problem into an ℓ_1 or ℓ_∞ optimization problem.

In this chapter, BCNs in the form of (2.32)–(2.33) are considered. For the sake of simplicity, the symbol $\ltimes$ will be omitted.

5.1 Trackability of reference output trajectory

In this section, a method to check the output trajectory trackability of BCN (2.32)–(2.33) will be proposed. For this purpose, let the output at time t be expressed as $y(t, x(0), \{u(0), u(1), \cdots, u(t-1)\}) = HLu(t-1)Lu(t-2)\cdots Lu(0)x(0)$. The reconstructibility of BCNs introduced in Fornasini and Valcher (2013) states that the current state $x(0)$ can be uniquely determined with the knowledge of any admissible input and output trajectory $\{(u(t), y(t)), t = -r_{min}, -r_{min}+1, \cdots, 0\}$ where r_{min} is the minimal reconstructibility index. Assume that the BCN (2.32)–(2.33) is reconstructible and the correct state estimate, i.e., $\hat{x}_s(0) = x(0)$, can be provided by the Luenberger-like observer (4.4) at time $t = 0$. The definition of output trajectory trackability is given as follows.

Definition 5.1 (Output Trajectory Trackability) Given a BCN (2.32)–(2.33) with the initial state $x(0)$ and a reference output trajectory from $t = 1$ to $t = T$, denoted

Z. Zhang, *Observer Design for Control and Fault Diagnosis of Boolean Networks*, https://doi.org/10.1007/978-3-658-35929-4_5

by $y^r(1), y^r(2), \cdots, y^r(T)$. The reference output trajectory is said to be trackable if there is a control sequence $U = \{u(0), u(1), \cdots, u(T-1)\}$ so that $y(t, x(0), U) = y^r(t), \forall t \in \{1, 2, \cdots, T\}$.

Assume that the reference output trajectory is given and denoted as $y^r(1) = \delta_{2^p}^{i_1}, y^r(2) = \delta_{2^p}^{i_2}, \cdots, y^r(T) = \delta_{2^p}^{i_T}$. For each output value $y^r(t) = \delta_{2^p}^{i_t}, t = 1, 2, \cdots, T$ a set $\Omega(t)$ and a vector $v(t)$ are defined, respectively, as

$$\Omega(t) = \left\{\delta_{2^n}^{i} | i \in \mathbb{Z}_+, Col_i(H) = \delta_{2^p}^{i_t}, 1 \le i \le 2^n\right\}, \tag{5.1}$$

$$v(t) = H^{\mathrm{T}} \delta_{2^p}^{i_t}. \tag{5.2}$$

It is noted that the set $\Omega(t)$ is called indistinguishability class in one step corresponding to the reference output $y^r(t) = \delta_{2^p}^{i_t}$ at time t (Fornasini and Valcher, 2013). The vector $v(t)$ is the vector expression of the set $\Omega(t)$. That is $\delta_{2^n}^{i} \in \Omega(t)$ if and only if $[v(t)]_i \neq 0$. If $\Omega(t) = \emptyset$ at some time t, then $v(t)$ is a zero vector and no state can generate the output $y^r(t)$.

Let a $2^n \times 2^n$ matrix M be defined as in Cheng et al. (2011):

$$M = L\mathbf{1}_{2^m}. \tag{5.3}$$

Lemma 5.2 Given a BCN (2.32)–(2.33) and a state $x(t) = \delta_{2^n}^{i_t}$. At least one state $x(t+1)$ in the indistinguishability class $\Omega(t+1)$ corresponding to the output $y^r(t+1)$ can be reached in one step if and only if

$$v(t+1)^{\mathrm{T}} M \delta_{2^n}^{i_t} \neq 0. \tag{5.4}$$

Based on Theorem 16.2 in Cheng et al. (2011) Lemma 5.2 can be proven. Therefore it is omitted.

Let $b(t+1) = v(t+1) \odot (M\delta_{2^n}^{i_t}) \in \mathbb{R}^{2^n \times 1}$. Lemma 5.2 tells us that starting from state $x(t) = \delta_{2^n}^{i_t}$, the state $x(t+1) = \delta_{2^n}^{j}$ that can generate the output $y^r(t+1)$ is

$$\begin{cases} \text{unreachable in one step, } & \text{if } [b(t+1)]_j = 0, \\ \text{reachable in one step, } & \text{if } [b(t+1)]_j > 0. \end{cases} \tag{5.5}$$

Hence, $\delta_{2^n}^{j} \in \Omega(t+1)$ is reachable in one step from $x(t) = \delta_{2^n}^{i_t}$ if and only if $[b(t+1)]_j > 0$.

Suppose that $x(0)$ is given as $x(0) = \delta_{2^n}^{i_0}$. Let $a(0) = x(0)$ and $a(t+1), t = 0, 1, \cdots, T-1$ be calculated forward as

$$a(t+1) = v(t+1) \odot (Ma(t)) \tag{5.6}$$

where the vector $v(t+1)$ is determined by (5.2) and the matrix M is defined in (5.3). Based on Lemma 5.2, the following result can be obtained.

Theorem 5.3 *The reference output trajectory* $\{y^r(1), y^r(2), \cdots, y^r(T)\}$ *is trackable by BCN (2.32)–(2.33) starting from* $x(0) = \delta_{2^n}^{i_0}$ *if and only if* $a(T) \neq \mathbf{0}_{2^n}$.

Proof According to Lemma 5.2 and (5.5), the set represented by vector $a(1)$ contains all possible states that are reachable in one step from the initial state $x(0) = \delta_{2^n}^{i_0}$ and generate the output $y^r(1)$. Similarly, if $[a(2)]_{i_2} > 0$, then there is one state $x(1) = \delta_{2^n}^{i_1}$ with $[a(1)]_{i_1} > 0$ so that $x(2) = \delta_{2^n}^{i_2}$ is reachable in one step from the state $x(1) = \delta_{2^n}^{i_1}$ and generates the output $y^r(2)$. Therefore, once $[a(t)]_{i_t} > 0$, there is at least one input sequence that steers the BCN (2.32)–(2.33) from the initial state $x(0) = \delta_{2^n}^{i_0}$ to the state $x(t) = \delta_{2^n}^{i_t}$ and generates the output trajectory $\{y^r(1), y^r(2), \cdots, y^r(t)\}$. If $a(t_1)$ is a zero vector, then, according to (5.6), the vectors $a(t_2), t_2 = t_1 + 1, \cdots, T$ are also zero vectors. Therefore, if and only if $a(T) \neq \mathbf{0}_{2^n}$, then the reference output trajectory is trackable. □

Remark 5.4 Assume that a given output trajectory is trackable starting from the state $\delta_{2^n}^{j}$. If a BCN is controllable, according to Cheng et al. (2011) it can be concluded that there is always an input sequence, such that any state $\delta_{2^n}^{i}$ can be driven to $\delta_{2^n}^{j}$. Hence, if a BCN is controllable and an output trajectory is trackable, then the BCN can always track the output trajectory with a time delay, no matter what the initial state $\delta_{2^n}^{i}$ is.

5.2 Exact tracking control

A case in which the given reference output trajectory is trackable will be considered. An approach will be given to find a control sequence which drives the BCN (2.32)–(2.33) starting from $x(0)$ to follow the trackable reference output trajectory $\{y^r(1), y^r(2), \cdots, y^r(T)\}$. The basic idea is to calculate $u(t)$ backwards based on the logical matrix L and the vectors $a(t), t = 0, 1, \cdots, T$ obtained by (5.6). Since the admissible state trajectory starting from $x(0)$ and generating the reference output trajectory may not be unique, our purpose is to find one control sequence that realizes an exact tracking. Denote the corresponding state trajectory as $\{x(0) = \delta_{2^n}^{i_0}, x(1) = \delta_{2^n}^{i_1}, \cdots, x(T) = \delta_{2^n}^{i_T}\}$. To achieve exact tracking control, the approach given in Algorithm 3 can be applied.

Algorithm 3: Given the BCN (2.32)–(2.33) with the known initial state $x(0) = \delta_{2^n}^{i_0}$ and a reference output trajectory $y^r(1), y^r(2), \cdots, y^r(T)$. Determine a control sequence $u^*(0), u^*(1), \cdots, u^*(T-1)$ that realizes exact tracking.

1. Check the output trajectory trackability and calculate $v(t)$ and $a(t)$, $t = 0, 1, \cdots, T$ by (5.2) and (5.6). Go to step 2 if the reference output trajectory is trackable.
2. Initialize $x(T) = \delta_{2^n}^{i_T}$ if $[a(T)]_{i_T} \neq 0$ and let $t = T - 1$.
3. Select an input $u^*(t) = \delta_{2^m}^{j}$ so that $[L_j a(t)]_{i_{t+1}} \neq 0$.
4. Select a state $x(t) = \delta_{2^n}^{i_t}$ so that $[a(t)]_{i_t} \neq 0$ and $[x^{\mathrm{T}}(t+1)L_j]_{i_t} \neq 0$.
5. If $t > 0$, replace t by $t - 1$ and return to Step 3.

Theorem 5.5 *If $[a(T)]_{i_T} \neq 0$, then the control sequence $\{u^*(0), u^*(1), \cdots, u^*(T-1)\}$ determined by Algorithm 3 steers the initial state $x(0) = \delta_{2^n}^{i_0}$ to $x(T) = \delta_{2^n}^{i_T}$ and generates the reference output trajectory $\{y^r(1), y^r(2), \cdots, y^r(T)\}$.*

Proof Initialize the state $x(T)$ as $\delta_{2^n}^{i_T}$ for the case of $[a(T)]_{i_T} \neq 0$. It is clear that $Hx(T) = y^r(T)$. If an input $u^*(T-1)$ is selected as $\delta_{2^m}^{j}$ so that $x^{\mathrm{T}}(T)Lu^*(T-1)a(T-1) \neq 0$, then there is at least one state $x(T-1)$ in the set represented by $a(T-1)$ that can be driven to the state $x(T)$ in one step (i.e. $L_j x(T-1) = x(T)$). Furthermore, if $[a(T-1)]_i \neq 0$, then one can obtain that $Hx(T-1) = H\delta_{2^n}^{i} = y^r(T-1)$. Repeating the same procedure, a control sequence $\{u^*(0), u^*(1), \cdots, u^*(T-1)\}$ is obtained. Applying the control sequence, the BCN (2.32)–(2.33) will generate the reference output trajectory $\{y^r(1), y^r(2), \cdots, y^r(T)\}$. □

Remark 5.6 Control problems for BCNs are in general $\mathcal{NP}$-hard (Akutsu et al., 2007). To analyze the computational complexity of Algorithm 3, consider the evaluation of the vectors $a(t)$ in (5.6). For each nonzero entry $[a(t-1)]_j \neq 0$, the entry of $v(t)$ corresponding to the canonical vector $Col_j(L_i)$ shall be compared with 0 for every index $i \in \{1, 2, \cdots, 2^m\}$. Let $K = \max_j(\mathbf{1}_{2^n}^{\mathrm{T}} \cdot Col_j(H^{\mathrm{T}})) \in [2^{n-p}, 2^n)$. The cardinality of the set satisfies $|\{i|[a(t)]_i \neq 0\}| \leq K$. The operation must be repeated for every $t = 1, 2, \cdots, T$. Hence, the computational burden to calculate the vectors $a(t), t = 1, 2, \cdots, T$ is $\mathcal{O}(T \cdot 2^m \cdot K)$. After that, in the worst case, considering steps 3 and 4 in Algorithm 3, the canonical column $Col_i(L_j)$ shall be selected for every index $j = 1, 2, \cdots, 2^m$ and every i so that $[a(t)]_i \neq 0$. Then, the i_{t+1}-th entry of $Col_i(L_j)$ will be compared with 0. The procedure will be repeated for every $t = T-1, T-2, \cdots, 0$. Since $a(0)$ contains only one nonzero entry, repeating the steps 3) and 4) requires $\mathcal{O}((T-1)\cdot 2^m \cdot K)$. As a result, Algorithm 3 runs in time at most $\mathcal{O}(T \cdot 2^m \cdot K)$.

5.3 Optimal tracking control

Now consider the case in which the reference output trajectory is not trackable. Then optimization problems can be formulated to determine an optimal control sequence that satisfies certain optimality criterion, for instance, that minimizes the total tracking error or the maximal tracking error over the given horizon.

5.3.1 Tracking error

Before the optimal tracking controller is designed, it is necessary to specify a measure of control performance. Recall that in coding theory, the concept of Hamming distance has been introduced by Richard Hamming in Hamming (1950) to measure the distance between two bit strings. Given two strings S and S^*, the Hamming distance is defined as the number of positions where S and S^* are different.

Deriving from by the Hamming distance, the distance between two outputs $Y(t)$ and $Y^r(t)$ denoted by $d_{out}(Y, Y^r, t)$ is defined as

$$d_{out}(Y, Y^r, t) = \sum_{i=1}^{p} \left(\left| Y_i(t) - Y_i^r(t) \right| \right) \tag{5.7}$$

where $|\cdot|$ represents the absolute value. Because there is a bijective correspondence between the two sets $\mathcal{D}^p$ and Δ_{2^p} (Fornasini and Valcher, 2013), the distance $d_{out}(Y, Y^r, t)$ between $Y(t)$ and $Y^r(t)$ at time t is considered as the tracking error $d(y, y^r, t)$.

As the reference output $y^r(t)$ is given, $d(y, y^r, t)$ is therefore a function of $y(t)$. If $y(t) = \delta_{2^p}^i$, then

$$\begin{aligned} d(y, y^r, t) &= d(\delta_{2^p}^i, y^r, t) \\ &= [d(\delta_{2^p}^1, y^r, t) \;\; d(\delta_{2^p}^2, y^r, t) \;\; \cdots \;\; d(\delta_{2^p}^{2^p}, y^r, t)]\delta_{2^p}^i \\ &= [d(\delta_{2^p}^1, y^r, t) \;\; d(\delta_{2^p}^2, y^r, t) \;\; \cdots \;\; d(\delta_{2^p}^{2^p}, y^r, t)]y(t). \end{aligned}$$

For the sake of simplicity, the function $d(y, y^r, t)$ is written as

$$d(y, y^r, t) = [d_1(t)\ d_2(t)\ \cdots\ d_{2^p}(t)]y(t), \tag{5.8}$$

$$d_i(t) = d(\delta_{2^p}^i, y^r, t),\ i = 1, 2, \cdots, 2^p \tag{5.9}$$

where $d_i(t)$ shows the tracking error with respect to the output $y(t) = \delta_{2^p}^i$. The vector $[d_1(t)\ d_2(t)\ \cdots\ d_{2^p}(t)]$ contains all possible output tracking errors at time t.

Notice that $\mathbf{1}_{2^m}^{\mathrm{T}} u(t) = 1$. Hence, (5.8) can be rewritten as

$$d(y, y^r, t) = [d_1(t)\ d_2(t)\ \cdots\ d_{2^p}(t)]\mathbf{1}_{2^m}^{\mathrm{T}} u(t)y(t). \tag{5.10}$$

The identity (5.10) will be used in the later derivation.

5.3.2 ℓ_1 optimization problem

In this subsection, the problem of minimizing the total tracking error over the horizon shall be considered. For this purpose, assume that a reference output trajectory $\{y^r(1), y^r(2), \cdots, y^r(T)\}$ is given. The total tracking error over the horizon can be evaluated by

$$e(x(0), y, y^r) = \sum_{t=1}^{T} d(y, y^r, t). \tag{5.11}$$

The ℓ_1 optimization problem is formulated as the problem of finding a control sequence that minimizes the total tracking error, i.e.,

$$\min_u J_1(x(0), u) = \min_u e(x(0), y, y^r). \tag{5.12}$$

Applying (5.10), the total tracking error $e(x(0), y, y^r)$ can be rewritten as

$$\begin{aligned} e(x(0), y, y^r) = {} & [d_1(T)\ d_2(T)\ \cdots\ d_{2^p}(T)]\, y(T) + \\ & \sum_{t=1}^{T-1} [d_1(t)\ d_2(t)\ \cdots\ d_{2^p}(t)]\, \mathbf{1}_{2^m}^{T} u(t)y(t). \end{aligned} \tag{5.13}$$

Let the non-negative weight factors c_f and $c(t)$, $t = 1, 2, \cdots, T-1$ be defined as

$$c_f = \begin{bmatrix} d_1(T) \\ d_2(T) \\ \vdots \\ d_{2^p}(T) \end{bmatrix} \text{ and } c(t) = \mathbf{1}_{2^m} \begin{bmatrix} d_1(t) \\ d_2(t) \\ \vdots \\ d_{2^p}(t) \end{bmatrix}. \tag{5.14}$$

Note that the weight factors c_f and $c(t)$ depend only on the reference output trajectory. The ℓ_1 performance index $J_1(x(0), u)$ can be equivalently written as

$$J_1(x(0), u) = c_f^{\mathrm{T}} y(T) + \sum_{t=1}^{T-1} c^{\mathrm{T}}(t) u(t) y(t). \tag{5.15}$$

According to Proposition 2.5 and (2.33), it holds that

$$\begin{aligned} J_1(x(0), u) &= c_f^{\mathrm{T}} H x(T) + \sum_{t=1}^{T-1} c^{\mathrm{T}}(t) u(t) H x(t) \qquad (5.16) \\ &= \left(c_f^{\mathrm{T}} H\right) x(T) + \sum_{t=1}^{T-1} c^{\mathrm{T}}(t) \left(I_{2^m} \otimes H\right) u(t) x(t) + \mathbf{0}_{2^{n+p}}^{\mathrm{T}} (I_{2^m} \otimes H) u(0) x(0). \end{aligned}$$

Let new weight factors $c_{f,new}$ and $c_{new}(t),\ t = 0, \cdots, T-1$ be

$$c_{f,new}^{\mathrm{T}} = c_f^{\mathrm{T}} H, \tag{5.17}$$

$$c_{new}^{\mathrm{T}}(t) = \begin{cases} \mathbf{0}_{2^{n+p}}^{\mathrm{T}} (I_{2^m} \otimes H) = \mathbf{0}_{2^{n+m}}^{\mathrm{T}}, & \text{if } t = 0, \\ c^{\mathrm{T}}(t)(I_{2^m} \otimes H), & \text{if } 1 \le t \le T-1. \end{cases}$$

Then the ℓ_1 optimization problem (5.12) is equivalent to

$$\begin{aligned} \min_u J_1(x(0), u) &= \min_u \left\{ c_{f,new}^{\mathrm{T}} x(T) + \sum_{t=0}^{T-1} c_{new}^{\mathrm{T}}(t) u(t) x(t) \right\}, \\ u^* &= \arg\min_u J_1(x(0), u). \end{aligned} \tag{5.18}$$

By looking at (5.18), it can be recognized that the ℓ_1 optimization problem can be solved with the algorithm derived in Fornasini and Valcher (2014b).

Theorem 5.7 (Fornasini and Valcher (2014b)) *Consider the optimization problem (5.18). The optimal control input u^* can be obtained by*

$$u^*(t) = K(t)x(t) \tag{5.19}$$

where the feedback gain matrix is

$$K(t) = \left[\delta_{2^m}^{i^*(1,t)}\ \delta_{2^m}^{i^*(2,t)}\ \cdots\ \delta_{2^m}^{i^*(2^n,t)}\right] \tag{5.20}$$

and the index $i^*(j,t)$, $j = 1, 2, \cdots, 2^n$ can be calculated according to Algorithm 4.

Algorithm 4: Given the BCN (2.32)-(2.33) and the weight factors $c_{new}(0)$, $c_{new}(1)$, $\cdots, c_{new}(T-1)$ and $c_{f,new}$. Determine $i^*(j,t)$, $j = 1, 2, \cdots, 2^n, t = 0, 1, \cdots, T-1$.

---

1. Partition $c_{new}(t)$ as

$$c_{new}(t) = [\tilde{\lambda}_1^{\mathrm{T}}(t)\ \tilde{\lambda}_2^{\mathrm{T}}(t)\ \cdots\ \tilde{\lambda}_{2^m}^{\mathrm{T}}(t)]^{\mathrm{T}}. \tag{5.21}$$

2. Initialize the vector $s(T) := c_{f,new}$.
3. Calculate the vector $s(t)$, $t = T-1, T-2, \cdots, 0$, element-wise ($j = 1, 2, \cdots, k$) and recursively:

$$[s(t)]_j := \min_{1 \le i \le 2^m} \left([\tilde{\lambda}_i(t)]_j + [s^{\mathrm{T}}(t+1)L_i]_j\right). \tag{5.22}$$

4. Calculate

$$i^*(j,t) = \arg \min_{1 \le i \le 2^m} \left([\tilde{\lambda}_i(t)]_j + [s^{\mathrm{T}}(t+1)L_i]_j\right). \tag{5.23}$$

Based on the initial state $x(0)$ and the feedback matrix $K(0)$, the input $u^*(0)$ can be obtained as $u^*(0) = K(0)x(0)$. Under the input $u^*(0)$, the initial state is transited to the state $x(1)$ according to (2.32). Thereby, the input $u^*(1)$ can be obtained. The procedure continues, until $t = T-1$. The ℓ_1 optimal tracking control approach is summarized in Algorithm 5.

Algorithm 5: Given the BCN (2.32)–(2.33) with known initial state $x(0) = \delta_{2^n}^{i_0}$ and a reference output trajectory $y^r(1), y^r(2), \cdots, y^r(T)$. Determine the optimal control sequence $u^*(t), t = 0, 1, \cdots, T-1$ that minimizes the total tracking error defined by (5.11).

1. Calculate the vectors $c_{f,new}$ and $c_{new}(t), t = 0, 1, \cdots, T-1$, by (5.9), (5.14) and (5.17).
2. Calculate the feedback matrix $K(t)$ backwards for $t = T-1, T-2, \cdots, 0$ according to (5.20).
3. Set $t = 0$.
4. Calculate $u^*(t)$ according to (5.19).
5. Calculate the state $x(t+1)$ according to (2.32).
6. If $t < T-1$, replace t by $t+1$ and return to Step 4.

It is worth noting that the solution to (5.23) in Algorithm 4 may not necessarily be unique (Fornasini and Valcher, 2014b). As they are all the feasible solutions, the index can be chosen arbitrarily. If a reference output trajectory is trackable, then the ℓ_1 optimization problem (5.12) has the minimum value zero and the optimal control sequence obtained by Algorithm 5 achieves exact tracking.

Remark 5.8 In order to analyze the computational complexity of the proposed algorithm, one should evaluate the distance determined by (5.7) at first which needs p comparisons and $p-1$ additions. So the computational complexity of calculating the distance (5.7) is $\mathcal{O}(p)$. This computation must be repeated for every output in the set Δ_{2^p}. In addition, this operation must be executed for every time $t = 1, 2, \cdots, T$. Hence, the computational burden required to calculate weight factors is $\mathcal{O}(p \cdot T \cdot 2^p)$. For solving the optimization problem (5.18), Fornasini and Valcher (2014b) showed that the computational complexity is $\mathcal{O}(T \cdot 2^{n+m})$.

Remark 5.9 The optimal tracking control can also be formulated as ℓ_2 optimization problem to minimize the total quadratic tracking error. Define the total quadratic tracking error as

$$\rho(y, y^r) = \sum_{t=1}^{T} d^2(y, y^r, t).$$

Then, applying the following weight factors for the ℓ_2 optimization problem

$$c_{2,f} = \begin{bmatrix} d_1^2(t_f) \\ d_2^2(t_f) \\ \vdots \\ d_{2^p}^2(t_f) \end{bmatrix} \text{ and } c_2(t) = \mathbf{1}_{2^m} \otimes \begin{bmatrix} d_1^2(t) \\ d_2^2(t) \\ \vdots \\ d_{2^p}^2(t) \end{bmatrix} \tag{5.24}$$

instead of the weight factors defined in (5.14), the same procedure as for the ℓ_1 optimization problem can be executed.

5.3.3 ℓ_∞ optimization problem

In order to avoid a large deviation during the tracking, an approach to solve the ℓ_∞ optimization problem is proposed which aims at minimizing the maximal tracking error.

Suppose that a reference output trajectory is given as $\{y^r(1), y^r(2), \cdots, y^r(T)\}$. Let the maximal tracking error over the horizon be represented by

$$\| d(y, y^r) \|_\infty = \max_{1 \le t \le T} d(y, y^r, t). \tag{5.25}$$

The ℓ_∞ optimization problem is formulated as

$$J_\infty^*(x(0)) = \min_u J_\infty(x(0), u) = \min_u \| d(y, y^r) \|_\infty . \tag{5.26}$$

By recalling (5.8), (5.10) and (5.14), the ℓ_∞ performance index $J_\infty(x(0), u)$ can be expressed as

$$\begin{aligned} J_\infty(x(0), u) &= \| d(y, y^r) \|_\infty \\ &= \max \left\{ d(y, y^r, T), \max_{1 \le t \le T-1} d(y, y^r, t) \right\} \\ &= \max \left\{ c_f^{\mathrm{T}} y(T), \max_{1 \le t \le T} c^{\mathrm{T}}(t) u(t) y(t) \right\} . \end{aligned} \tag{5.27}$$

Notice that $\| d(y, y^r) \|_\infty$ is always non-negative. Applying the weight factors $c_{f,new}$ and $c_{new}(t)$ defined in (5.17) and $c_{new}^{\mathrm{T}}(0)u(0)x(0) = 0$, the ℓ_∞ performance index $J_\infty(x(0), u)$ can be equivalently rewritten as

$$J_\infty(x(0), u) = \max \left\{ c_{f,new}^{\mathrm{T}} x(T), \max_{0 \le t \le T-1} c_{new}^{\mathrm{T}}(t) u(t) x(t) \right\} \tag{5.28}$$

which has a form similar to the ℓ_1 performance index $J_1(x(0), u)$ in (5.15) with additions replaced by max-operations. Hence, recalling (5.21) and applying the Bellman's Principle of Optimality, equations similar to (5.22) and (5.23) can be derived which were originally given in Fornasini and Valcher (2014b).

Theorem 5.10 *Consider the ℓ_∞ optimization problem (5.26). The vector $w(T)$ is initialized as $c_{f,new}$ and the vectors $w(t)$, $t = T-1, T-2, \cdots, 0$ are determined element-wise by*

$$[w(t)]_j = \min_{1 \leq i \leq 2^m} \max\{[\tilde{\lambda}_i(t)]_j, [w^\mathrm{T}(t+1)L_i]_j\}. \tag{5.29}$$

The optimal control input can be obtained as (5.19) and (5.20) where the index $i^*(j,t)$, $j = 1, 2, \cdots, 2^n$ is determined by

$$i^*(j,t) = \arg \min_{1 \leq i \leq 2^m} \max\{[\tilde{\lambda}_i(t)]_j, [w^\mathrm{T}(t+1)L_i]_j\}. \tag{5.30}$$

According to (2.32) and based on $K(t)$ and the initial state $x(0)$, the optimal input sequence $u^*(t)$, $t = 0, 1, \cdots, T-1$ can be calculated. Similar to the ℓ_1 optimization problem introduced in Section 5.3.2, the solution to (5.30) in Theorem 5.10 may not necessarily be unique. As they are all the feasible solutions, the index can be chosen arbitrarily.

Remark 5.11 If a reference trajectory is trackable, then the ℓ_∞ optimization problem (5.26) has the minimum value zero. The optimal control sequence obtained by applying Theorem 5.10 achieves exact tracking.

5.3.4 Penalty for changes in control inputs

In some situations, frequent changes in control inputs are not desired. As mentioned in Kauffman (1969), Boolean networks can be used to model gene regulatory network which can be applied for the treatment of diseases such as cancer (Faryabi et al., 2008). For this, drug delivery in disease treatment will be taken as an example. Usually, drugs remain active in the body for a particular length of time (the duration of activation) which may be indicated by their biological half life. The biological half life is the time required to lose 50% of the pharmacological, physiologic or radiologic activity of a substance (Toutain and BOUSQUET-MÉLOU, 2004). This may vary from a couple of hours to longer than a day. Moreover, it is infeasible to change the drugs applied to a patient several times within a very short time period. In this section, it is shown that the ℓ_1 and ℓ_∞ optimal tracking control proposed above can be modified to take the cost for changes in control inputs into account.

Changes in control inputs can be defined in a similar way as the tracking error introduced in Subsection 5.3.1. Motivated by the Hamming distance, the distance between two inputs $U(t)$ and $U(t-1)$, denoted by $d_{in}(U, Y^r, t)$, is defined as

$$d_{in}(U(t), U(t-1)) = \sum_{i=1}^{m} |U_i(t) - U_i(t-1)| \tag{5.31}$$

where $|\cdot|$ represents the absolute value. Since there is a bijective correspondence between the two sets $\mathcal{D}^m$ and Δ_{2^m} (Fornasini and Valcher, 2013), the distance $d_{in}(U(t), U(t-1)$ between $U(t)$ and $U(t-1)$ is considered as the change of control inputs $\zeta(u(t), u(t-1))$, i.e., $\zeta(u(t), u(t-1)) = d_{in}(U(t), U(t-1))$.

For the given inputs $u(t)$ and $u(t-1)$, the changes in control inputs $\zeta(u(t), u(t-1))$ are uniquely determined. Therefore, $\zeta(u(t), u(t-1))$ is a function of $u(t)$ and $u(t-1)$. As mentioned in Fornasini and Valcher (2014b), the function $\zeta(u(t), u(t-1))$ can be represented in an algebraic form as

$$\zeta(u(t), u(t-1)) = D_{in}u(t)u(t-1) \tag{5.32}$$

where

$$D_{in} = [\zeta(\delta_{2^m}^1, \delta_{2^m}^1)\ \zeta(\delta_{2^m}^1, \delta_{2^m}^2) \cdots \zeta(\delta_{2^m}^1, \delta_{2^m}^{2^m})\ \zeta(\delta_{2^m}^2, \delta_{2^m}^1) \cdots \zeta(\delta_{2^m}^{2^m}, \delta_{2^m}^{2^m})] \tag{5.33}$$

contains penalty for all possible changes in control inputs. The total changes of the control inputs over the horizon are defined as

$$g(u) = \sum_{t=0}^{T-1} \zeta(u(t), u(t-1)) = \sum_{t=0}^{T-1} D_{in}u(t)u(t-1). \tag{5.34}$$

In order to consider the penalty for changes in control inputs, the goal of the ℓ_1 optimization problem is now to find a control sequence that minimizes the weighted sum of the total tracking error and the total changes in control inputs, i.e.,

$$\min_u J_{1,\Delta}(x(0), u) = \min_u \left\{(1-\beta) \cdot e(x(0), y, y^r)+\beta \cdot g(u)\right\} \tag{5.35}$$

where $0 \leq \beta < 1$ is the weight factor.

Recalling (5.13), (5.17) and (5.34), the cost function $J_{1,\Delta}(x(0), u)$ can be expressed as

$$J_{1,\Delta}(x(0), u) = (1-\beta) \cdot \left(c_{f,new}^{\mathrm{T}} x(T) + \sum_{t=0}^{T-1} c_{new}^{\mathrm{T}}(t)u(t)x(t)\right) + \beta \cdot \sum_{t=0}^{T-1} D_{in}u(t)u(t-1) \tag{5.36}$$

which, in fact, has a form similar to the cost function introduced as Section V in Fornasini and Valcher (2014b). As already shown in Fornasini and Valcher (2014b), to solve the optimization problem (5.35), augmented state variables

$$\tilde{x}(t) = x(t)u(t-1), t = 0, 1, \cdots, T \tag{5.37}$$

are introduced and the weight factors π_f^{T} and $\pi^{\mathrm{T}}(t), t = 0, 1, \cdots, T-1$ are defined as

$$\begin{aligned} \pi_f^{\mathrm{T}} &= (1-\beta) \cdot \left(c_{f,new}^{\mathrm{T}} \otimes \mathbf{1}_{2^m}^{\mathrm{T}}\right), \\ \pi^{\mathrm{T}}(t) &= (1-\beta) \cdot \left(c_{new}^{\mathrm{T}}(t) \otimes \mathbf{1}_{2^m}^{\mathrm{T}}\right) + \beta \cdot \left(\mathbf{1}_{2^n}^{\mathrm{T}} \otimes D_{in}^{\mathrm{T}}\right) W_{[2^m,2^n]}. \end{aligned} \tag{5.38}$$

Based on this, the cost function (5.36) can be equivalently rewritten as

$$J_{1,\Delta}(x(0), u) = \pi_f^{\mathrm{T}} \tilde{x}(T) + \sum_{t=0}^{T-1} \pi^{\mathrm{T}}(t)u(t)\tilde{x}(t). \tag{5.39}$$

Furthermore, according to (2.32), there is

$$\begin{aligned} \tilde{x}(t+1) &= x(t+1)u(t) \\ &= Lu(t)x(t)u(t) \\ &= Lu(t)W_{[2^m,2^n]}u(t)x(t)\mathbf{1}_{2^m}^{\mathrm{T}}u(t-1) \\ &= L(I_{2^m} \otimes W_{[2^m,2^n]})\Phi_{2^m}u(t)x(t)\mathbf{1}_{2^m}^{\mathrm{T}}u(t-1) \\ &= (L(I_{2^m} \otimes W_{[2^m,2^n]})\Phi_{2^m}) \otimes \mathbf{1}_{2^m}^{\mathrm{T}}u(t)x(t)u(t-1). \end{aligned} \tag{5.40}$$

Let $\tilde{L} = (L(I_{2^m} \otimes W_{[2^m,2^n]})\Phi_{2^m}) \otimes \mathbf{1}_{2^m}^{\mathrm{T}}$. So the transition of the variable $\tilde{x}(t)$ can be described by

$$\tilde{x}(t+1) = \tilde{L}u(t)\tilde{x}(t) \tag{5.41}$$

which has a form similar to (2.32). By regarding (5.37) and the weight factors defined in (5.38), the optimization problem (5.35) has a form similar to the ℓ_1 performance index $J_1(x(0), u)$ and can thereby be solved by applying Theorem 5.7.

In the next step, it will be shown how the cost for changes in control inputs can be considered in the ℓ_∞ optimization problem. The ℓ_∞ optimization problem is formulated as the problem of finding a control sequence that minimizes the maximum of the weighted sum of the tracking error and the cost for changes in control inputs, i.e.

$$\begin{aligned} \min_u J_{\infty,\Delta}(x(0), u) = \min_u \Bigg\{ \max \Big\{ & (1-\beta) \cdot d(y, y^r, T), \\ & \max_{1 \le t \le T-1} \{(1-\beta) \cdot d(y, y^r, t) + \beta \cdot \varsigma(u(t), u(t-1))\} \Big\} \Bigg\} \end{aligned} \tag{5.42}$$

where $0 \le \beta < 1$ is the weight factor.

Recalling (5.28) and (5.32), the cost function $J_{\infty,\Delta}(x(0), u)$ for the ℓ_∞ optimization problem (5.42) can be written as

$$\begin{aligned} J_{\infty,\Delta}(x(0), u) = \max \Big\{ & (1-\beta) \cdot c_{f,new}^{\mathrm{T}} x(T), \max_{0 \le t \le T-1} \Big\{ \\ & (1-\beta) \cdot c_{new}^{\mathrm{T}}(t) u(t) x(t) + \beta \cdot D_{in} u(t) u(t-1) \Big\} \Big\}. \end{aligned} \tag{5.43}$$

Applying the weight factors π_f^{T} and $\pi^{\mathrm{T}}(t)$, $t = 0, 1, \cdots, T-1$ defined in (5.38) and the state variable defined in (5.37), the cost function $J_{\infty,\Delta}(x(0), u)$ can be equivalently rewritten as

$$J_{\infty,\Delta}(x(0), u) = \max \left\{ \pi_f^{\mathrm{T}} \tilde{x}(T), \max_{0 \le t \le T-1} \left\{ \pi^{\mathrm{T}}(t) u(t) \tilde{x}(t) \right\} \right\}. \tag{5.44}$$

Therefore, by recalling (5.41), Theorem 5.10 can be applied to solve the optimization problem (5.42).

5.4 Handling of constraints

In real systems, the controller design may be subject to constraints. For instance, during treatment of diseases, some states of genes must be avoided because they may lead to other diseases (Chen et al., 2015). Moreover, side effects could be caused either by the use of a certain medicine under some states of genes or by some drug combinations. One advantage of the approaches proposed in Section 5.2 and 5.3 is that such kinds of state, transition and input constraints can be easily taken into account in the design procedure.

For the trackability analysis and design of an exact tracking controller, the basic idea in handling constraints is to delete the corresponding transitions. For the ℓ_1 and ℓ_∞ optimization problems, such as (5.12) and (5.26), the final state $x(T)$ and the state $x(t)$, together with input $u(t)$, are weighted, respectively, by the weight factors $c_{f,new}$ and $c_{new}(t)$ in the cost function. The forbidden states, transitions and inputs can be punished by imposing the corresponding coefficients in the cost function as ∞.

5.4.1 State constraints

Without loss of generality, assume that the state $\delta_{2^n}^i$ should be avoided. In the trackability analysis and exact tracking control, the i-th row of the matrix M must be set to

$$Row_i(M) = 0. \tag{5.45}$$

By doing this modification, any transition from other states to the state $\delta_{2^n}^i$ will not be considered.

In the case of the ℓ_1 and ℓ_∞ optimization problems (5.18) and (5.26), the weight factors $c_{f,new}$ and $c_{new}(t)$ need to be modified. Because $x(T) = \delta_{2^n}^i$ is weighted by $[c_{f,new}]_i$ (i.e. $c_{f,new}^{\mathrm{T}} \delta_{2^n}^i = [c_{f,new}]_i$), the i-th entry of the vector $c_{f,new}$ needs to be changed to ∞. Furthermore, to avoid state $x(t) = \delta_{2^n}^i$, $t = 0, 1, \cdots, T-1$, the vector $c_{new}(t)$ will be partitioned into 2^m blocks as

$$c_{new}(t) = [\tilde{\lambda}_1^{\mathrm{T}}(t)\ \tilde{\lambda}_2^{\mathrm{T}}(t)\ \cdots\ \tilde{\lambda}_{2^m}^{\mathrm{T}}(t)]^{\mathrm{T}}.$$

Each block $\tilde{\lambda}_j(t)$ is modified as

$$\left[\lambda\tilde{(t)}_j\right]_i = \infty,\ j = 1, 2, \cdots, 2^m;\ t = 0, 1, \cdots, T-1. \tag{5.46}$$

As any state constraint does not influence the penalty for changes in control inputs, the same method to modify the weight factors $c_{f,new}$ and c_{new} can be applied. Based on this, the weight factors π_f and $\pi(t)$ are calculated according to (5.38) for the ℓ_1 and ℓ_∞ optimization problems (5.35) and (5.42). Consequently, it can be seen that the states $\delta_{2^n}^i$ can be avoided successfully if the optimal performance J_1^* and J_∞^* are finite.

5.4.2 Transition constraints

In the framework of BCNs, a forbidden transition means the forbidden use of input $u(t)$ in combination with a certain state $x(t)$. For instance, the input $u(t) = \delta_{2^m}^j$ is not allowed if the state is $x(t) = \delta_{2^n}^i$. A simple computation shows that $u(t)x(t) = \delta_{2^m}^j \delta_{2^n}^i = \delta_{2^{m+n}}^{2^n(j-1)+i}$. According to (2.32), the $(2^n(j-1)+i)$-th column of the logical matrix L corresponds to the results of the transition. For the trackability analysis and the exact tracking control, L should be modified as follows:

$$Col_{2^n(j-1)+i}(L) = \mathbf{0}_{2^n}. \tag{5.47}$$

According to (5.3), if L is changed, then the matrix M changes correspondingly.

To consider the forbidden transition during the design of the optimal tracking controller without considering the cost for the change in control inputs, state $x(t)$, together with input $u(t)$, is weighted by the weight factors $c_{new}(t)$, i.e. $c_{new}^{\mathrm{T}}(t)u(t)x(t)$. If $u(t) = \delta_{2^m}^j$ and $x(t) = \delta_{2^n}^i$, then there is $c_{new}^{\mathrm{T}}(t)u(t)x(t) = c_{new}^{\mathrm{T}}(t)\delta_{2^m}^j\delta_{2^n}^i = [c_{new}^{\mathrm{T}}(t)]_{(j-1)\cdot 2^n+i}$. In order to avoid the transition, the $(j-1)\cdot 2^n + i$-th entry of the weight factor $c_{new}^{\mathrm{T}}(t)$ needs to be modified as

$$[c_{new}^{\mathrm{T}}(t)]_{(j-1)\cdot 2^n+i} = \infty,\ t = 0, 1, \cdots, T-1. \tag{5.48}$$

For the ℓ_1 and ℓ_∞ optimization problems (5.35) and (5.42), state $x(t)$, inputs $u(t-1)$ and $u(t)$ are weighted by $\pi(t)$ (i.e. $\pi^{\mathrm{T}}(t)u(t)x(t)u(t-1)$). Partition the weight factor $\pi(t)$ into 2^{m+n} blocks, i.e.

$$\pi^{\mathrm{T}}(t) = [Blk_1(\pi^{\mathrm{T}}(t))\ \ Blk_2(\pi^{\mathrm{T}}(t))\ \ \cdots\ \ Blk_{2^{n+m}}(\pi^{\mathrm{T}}(t))]. \tag{5.49}$$

If $u(t) = \delta_{2^m}^j$ and $x(t) = \delta_{2^n}^i$, then it is clear that

$$\pi^{\mathrm{T}}(t)u(t)x(t)u(t-1) = Blk_{(j-1)\cdot 2^n+i}(\pi^{\mathrm{T}}(t))u(t-1).$$

Hence, in order to consider the transition constraints, the $(j-1)\cdot 2^n + i$-th block needs to be changed to

$$Blk_{(j-1)\cdot 2^n+i}(\pi^{\mathrm{T}}(t)) = \infty \otimes \mathbf{1}^{\mathrm{T}}_{2^m},\ t = 0, 1, \cdots, T-1. \tag{5.50}$$

5.4.3 Input constraints

Assume that the input $\delta^j_{2^m}$ is not realizable and thus must be avoided. According to (2.32), any result of transitions associated with the input $u(t) = \delta^j_{2^m}$ can be found in the matrix $L_j = L\delta^j_{2^m}$. In case of trackability analysis and exact tracking control, the logical matrix L is split into 2^m blocks of the dimensions $2^n \times 2^n$. In the next step, all entries in L_j must be set to 0. With the changed logical matrix L, the matrix M is modified accordingly.

To handle input constraints in the optimization problems (5.12) and (5.26), input $u(t)$ is weighted by the weight factors $c_{new}(t)$, i.e. $c^{\mathrm{T}}_{new}(t)u(t)x(t)$. The weight factors $c_{new}(t)$ are partitioned as

$$c^{\mathrm{T}}_{new}(t) = [\tilde{\lambda}^{\mathrm{T}}_1(t)\ \tilde{\lambda}^{\mathrm{T}}_2(t)\ \cdots\ \tilde{\lambda}^{\mathrm{T}}_{2^m}(t)]^{\mathrm{T}}. \tag{5.51}$$

Because of $\tilde{\lambda}_j(t) = (c^{\mathrm{T}}_{new}(t) \ltimes \delta^j_{2^m})^{\mathrm{T}}$, $\tilde{\lambda}_j(t)$ must be set to

$$\tilde{\lambda}_j(t) = \infty \cdot (\mathbf{1}^{\mathrm{T}}_{2^n} \otimes \mathbf{1}_{2^n}),\ \ t = 0, 1, \cdots, T-1. \tag{5.52}$$

For the optimization problems with penalty for changes in control inputs (5.35) and (5.42), input $u(t),\ t = 0, 1, \cdots, T-1$ is weighted by the weight factors $\pi(t)$. Therefore, let $\pi(t)$ be split into 2^m blocks as

$$\pi^{\mathrm{T}}(t) = [Blk_1(\pi^{\mathrm{T}}(t))\ \ Blk_2(\pi^{\mathrm{T}}(t))\ \ \cdots\ \ Blk_{2^m}(\pi^{\mathrm{T}}(t))]. \tag{5.53}$$

Then the j-th block of $\pi(t),\ t = 0, 1, \cdots, T-1$ should be changed as

$$Blk_j(\pi(t)) = \infty \otimes \mathbf{1}^{\mathrm{T}}_{2^{m+n}},\ t = 0, 1, \cdots, T-1. \tag{5.54}$$

5.5 Example

In order to illustrate the main results in this chapter, consider the following BCN:

$$\begin{cases} X_1(t+1) = \neg U_1(t) \wedge (X_2(t) \vee X_3(t)), \\ X_2(t+1) = \neg U_1(t) \wedge U_2(t) \wedge X_1(t), \\ X_3(t+1) = \neg U_1(t) \wedge (U_2(t) \vee (U_3(t) \wedge X_1(t))), \\ Y_1(t) = X_1(t), \\ Y_2(t) = X_2(t). \end{cases} \tag{5.55}$$

which is a reduced BCN model for the *lac* operon in the bacterium *Escherichia coli* derived in Veliz-Cuba and Stigler (2011). In the model, X_1 represents transcription of mRNA. The states X_2 and X_3 indicate, respectively, a high and a medium concentration of lactose. U_1 represents an abundance of extra cellular glucose. U_2 and U_3 denote, respectively, a high and a medium concentration of extra cellular lactose. Assume that the states X_1 and X_2 can be measured, X_3 is not measured.

Using the semi-tensor product of matrices, (5.55) can be converted into the following equivalent form:

$$\begin{cases} x(t+1) = Lu(t)x(t), \\ y(t) = Hx(t) \end{cases} \tag{5.56}$$

Table 5.1 Reference output trajectory

(a) Output trajectory 1

t	1	2	3	4
Y_1^r	1	0	0	0
Y_2^r	0	1	0	0
y^r	δ_4^2	δ_4^3	δ_4^4	δ_4^4

(b) Output trajectory 2

t	1	2	3	4
Y_1^r	0	1	0	0
Y_2^r	1	0	0	0
y^r	δ_4^3	δ_4^2	δ_4^4	δ_4^4

where $x(t) = \ltimes_{i=1}^3 x_i(t) \in \Delta_8$, $u(t) = \ltimes_{i=1}^3 u_i(t) \in \Delta_8$, $y(t) = \ltimes_{i=1}^2 y_i(t) \in \Delta_4$ and

$$\begin{aligned} L = {} & [L_1 \mid L_2 \mid L_3 \mid L_4 \mid L_5 \mid L_6 \mid L_7 \mid L_8] \\ = {} & \delta_8[8\ 8\ 8\ 8\ 8\ 8\ 8\ 8 \mid 8\ 8\ 8\ 8\ 8\ 8\ 8\ 8 \mid 8\ 8\ 8\ 8\ 8\ 8\ 8\ 8 \mid 8\ 8\ 8\ 8\ 8\ 8\ 8\ 8 \\ & 1\ 1\ 1\ 5\ 3\ 3\ 3\ 7 \mid 1\ 1\ 1\ 5\ 3\ 3\ 3\ 7 \mid 3\ 3\ 3\ 7\ 4\ 4\ 4\ 8 \mid 4\ 4\ 4\ 8\ 4\ 4\ 4\ 8], \\ H = {} & \delta_4[1\ 1\ 2\ 2\ 3\ 3\ 4\ 4]. \end{aligned} \tag{5.57}$$

Assume that $x(0) = \delta_8^2$. Two reference output trajectories are given in Table 5.1a and Table 5.1b, respectively.

Suppose that the state δ_8^1 must be avoided as it represents a state constraint. In addition, if the system is in the state $x(t) = \delta_8^6$, then the use of the input $u(t) = \delta_8^3, \forall t \in \mathbb{Z}_+$ is not allowed which is indeed a transition constraint. Further, according to the specification in Veliz-Cuba and Stigler (2011), (u_2, u_3) can only take (0, 0), (0, 1) and (1, 1) values to represent, respectively, the low, medium and high concentration of extra cellular lactose. Therefore, the inputs δ_8^2 and δ_8^6 must be avoided due to lack of physical meaning which describes the input constraints.

At first, the reference output trajectory 1 is considered. In the trackability analysis, the matrix M is obtained by (5.3) and then modified according to (5.45) and (5.47). As a result,

$$M = \begin{bmatrix} 0 & 0 & 0 & 0 & 0 & 0 & 0 & 0 \\ 0 & 0 & 0 & 0 & 0 & 0 & 0 & 0 \\ 1 & 1 & 1 & 0 & 1 & 1 & 1 & 0 \\ 1 & 1 & 1 & 0 & 2 & 2 & 2 & 0 \\ 0 & 0 & 0 & 1 & 0 & 0 & 0 & 0 \\ 0 & 0 & 0 & 0 & 0 & 0 & 0 & 0 \\ 0 & 0 & 0 & 1 & 0 & 0 & 0 & 1 \\ 3 & 3 & 3 & 4 & 3 & 2 & 3 & 5 \end{bmatrix}. \tag{5.58}$$

Initializing $a(0) = x(0) = \delta_8^2$, the vectors $a(t), t = 1, 2, 3, 4$ are determined according to (5.6) as

$$\begin{aligned} a(1) &= [0\,0\,1\,1\,0\,0\,0\,0]^{\mathrm{T}}, \\ a(2) &= [0\,0\,0\,0\,1\,0\,0\,0]^{\mathrm{T}}, \\ a(3) &= [0\,0\,0\,0\,0\,0\,0\,3]^{\mathrm{T}}, \\ a(4) &= [0\,0\,0\,0\,0\,0\,3\,15]^{\mathrm{T}}. \end{aligned}$$

As $a(4) \neq 0$, the reference output trajectory 1 is trackable under the state, transition and input constraints. After execution of Algorithm 3, the following input sequence that realizes exact tracking can be obtained:

$$\begin{aligned} u(0) &= \delta_8^8, \\ u(1) &= \delta_8^5, \\ u(2) &= \delta_8^3, \\ u(3) &= \delta_8^1. \end{aligned} \tag{5.59}$$

In the same way, for reference output trajectory 2 $a(1) = a(2) = a(3) = a(4) = \mathbf{0}_8$ is obtained. Hence, an exact tracking of reference output trajectory 2 under the given constraints is not possible. Assume that the total tracking error should be minimized. For this purpose, the ℓ_1 optimization approach is applied. The weight factors $c_{f,new}$ and $c_{new}(t)$, $t = 0, 1, 2, 3$ are calculated by (5.17) and then modified. For instance, the vector $c_{new}(3)$ is

$$\begin{matrix}[\infty & 2 & 1 & 1 & 1 & 1 & 0 & 0 & \infty & \infty & \infty & \infty & \infty & \infty & \infty & \infty \\ \infty & 2 & 1 & 1 & 1 & \infty & 0 & 0 & \infty & 2 & 1 & 1 & 1 & 1 & 0 & 0 \\ \infty & 2 & 1 & 1 & 1 & 1 & 0 & 0 & \infty & \infty & \infty & \infty & \infty & \infty & \infty & \infty \\ \infty & 2 & 1 & 1 & 1 & 1 & 0 & 0 & \infty & 2 & 1 & 1 & 1 & 1 & 0 & 0]^{\mathrm{T}}.\end{matrix}$$

After that, the ℓ_1 optimization problem (5.18) is solved which yields $u^*(0) = \delta_8^1$, $u^*(1) = \delta_8^1$, $u^*(2) = \delta_8^1$, $u^*(3) = \delta_8^1$. The minimal total tracking error is $J_1^*(x(0), u) = J_1^*(\delta_8^2, u^*) = 1$.

Model-Based Fault Diagnosis

6

In this chapter, passive and active fault diagnosis problems of BCNs will be investigated. At first, passive fault diagnosis problem of BCNs is studied. For this purpose, a necessary and sufficient condition for the fault detectability is proposed. Then, a reduced-order observer is proposed for fault detection of BCNs. If fault occurrence has been detected, then an approach is proposed to solve fault isolation problem of BCNs. The basic idea is to separate dynamics of the faults into different independent subsystems by using graph theory. For each subsystem, a residual generator is constructed based on reduced-order observer by getting rid of indistinguishable states. After that, motivated by Fiacchini and Millérioux (2019) an approach is introduced to solve active fault diagnosis problem. It is shown that the approach requires substantially lower computational complexity than the approach proposed in Fornasini and Valcher (2015b).

The BCNs considered in this chapter are described in the form (2.34)–(2.35) and the Boolean product (2.36) will be applied. For the sake of simplicity, the symbol $\ltimes_{\mathcal{B}}$ will be omitted. In order to distinguish faulty BCN from non-faulty BCN, let the subscript "F" explicitly indicate a faulty behavior and the following algebraic form is introduced to represent a faulty BCN

$$x_F(t+1) = L_F x_F(t) u(t), \tag{6.1}$$

$$y_F(t) = H_F x_F(t), \tag{6.2}$$

where $x_F(t) \in \Delta_{2^n}$, $y_F(t) \in \Delta_{2^p}$ are, respectively, the state and output under the effect of fault, $L_F \in \mathcal{L}_{2^n \times 2^{n+m}}$, $H_F \in \mathcal{L}_{2^p \times 2^n}$ are the logical matrices of faulty BCN. It is important to note that the faulty BCN (6.1)–(6.2) is a more general faulty model than the one described in Fornasini and Valcher (2015a), where fault can only

Z. Zhang, *Observer Design for Control and Fault Diagnosis of Boolean Networks*,
https://doi.org/10.1007/978-3-658-35929-4_6

occur in state equation (6.1). In the same way, in order to study the fault diagnosis problem of BCNs, the following algebraic form is introduced to model the j-th fault:

$$x_{F_j}(t+1) = L_{F_j} x_{F_j}(t) u(t), \tag{6.3}$$

$$y(t) = H_{F_j} x_{F_j}(t), \tag{6.4}$$

where L_{F_j} and H_{F_j} are, respectively, logical matrices of dimensions $2^n \times 2^{n+m}$ and $2^p \times 2^n$ that describe the dynamic of the j-th faulty system. In the same way as mentioned in Fornasini and Valcher (2015a), it is assumed that a BCN cannot recover itself autonomously after a fault occurs.

6.1 Passive Fault Diagnosis

In this section, the passive fault detection will be studied.

6.1.1 Observability Analysis

The property "observability" is introduced to measure whether the initial state of the BCNs can be inferred with the knowledge of the input and output trajectory. In this section, an efficient algorithm for the observability analysis of BCNs will be proposed. The basic idea is to convert the observability analysis problem of BCNs to a relational coarsest partition problem (RCP) of transition systems. In the following part, the observability defined in Zhao et al. (2010) is considered.

Definition 6.1 (Observability (Zhao et al., 2010)) A BCN is said to be observable if for any two distinct states $\bar{x}(0)$, $x(0)$ there is an input sequence $\{u(0), u(1), \cdots\}$, such that the corresponding output sequences $\{\bar{y}(0), \bar{y}(1), \cdots\} \neq \{y(0), y(1), \cdots\}$.

The RCP problem decides the equivalences of states in a labeled transition system (Fernandez, 1990). The observability defined in Definition 6.1 shows that all the states are inequivalent as at least one generated input and output trajectories do not coincide. Hence, the RCP problem can be applied to observability analysis. According to Fernandez (1990), a labeled transition system is a quadruple $S = (\mathcal{Q}, \mathcal{A}, \mathcal{T}, q(0))$ where $\mathcal{Q}$ is a finite set of states, $\mathcal{A}$ is a finite set of actions, $\mathcal{T} \subseteq \mathcal{Q} \times \mathcal{A} \times \mathcal{Q}$ is the set of transition relations and $q(0)$ is the initial state.

With respect to an action $a \in \mathcal{A}$, the transition relation $T_a(q)$ is defined as

$$T_a(q) = \{q' | (q, a, q') \in \mathcal{T}\} \tag{6.5}$$

where (q, a, q') means that q' is a successor state of q by applying the action a, i.e. $q \xrightarrow{a} q'$. It is important to note that when applying an action a, the state q can be steered to more than one successor state in the set $\mathcal{Q}$. In a similar way, the states, that can be transient to q' by using a, can be denoted as $T_a^{-1}(q') = \{q | (q, a, q') \in \mathcal{T}\}$.

A partition of a set is to group the elements of a set into subsets so that each element is included in only one subset. Let $\rho = \{B_1, B_2, \cdots, B_{|\rho|}\}$ and $\rho' = \{B_1', B_2', \cdots, B_{|\rho'|}'\}$ be two different partitions of the state space $\mathcal{Q}$ which is a finite set. According to Fernandez (1990), the partition ρ' is called a refinement of the partition ρ (or the partition ρ is coarser than the partition ρ') if and only if $\forall i = 1, 2, \cdots, |\rho'|$ there is an index $j \in \{1, 2, \cdots, |\rho|\}$ so that $B_i' \subseteq B_j$.

Definition 6.2 (Fernandez (1990)) A partition ρ of the state space $\mathcal{Q}$ is compatible with the binary relation T_a if and only if for any $i, j \in \{1, 2, \cdots, |\rho|\}$ it holds that

$$\forall q, \tilde{q} \in B_i, T_a(q) \cap B_j \neq \emptyset \Leftrightarrow T_a(\tilde{q}) \cap B_j \neq \emptyset. \tag{6.6}$$

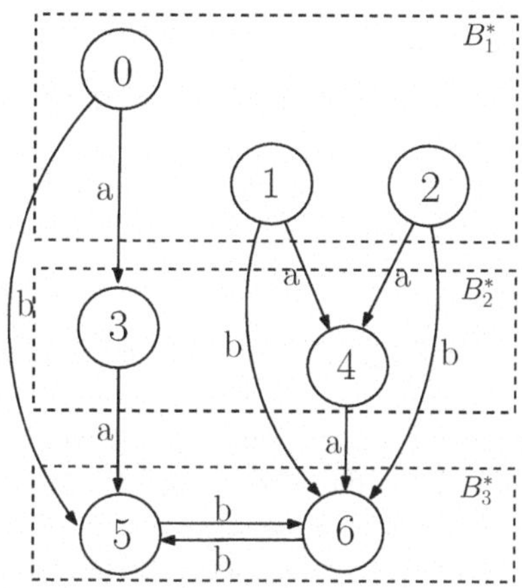

Figure 6.1 A labeled transition system

If the condition (6.6) is satisfied for all possible actions $a \in A$, then the partition ρ is compatible with the transition relations $\mathcal{T}$. Based on this, the RCP problem is defined as follows.

Definition 6.3 (RCP problem (Fernandez, 1990)) Given a partition ρ of the state space $\mathcal{Q}$ and a transition relation $\mathcal{T} \subseteq \mathcal{Q} \times \mathcal{A} \times \mathcal{Q}$. The relational coarsest partition problem is to find the coarsest refinement ρ^* of the partition ρ so that ρ^* is compatible with the transition relation $\mathcal{T}$.

To solve the RCP problem, an efficient algorithm has been proposed in Paige and Tarjan (1987). Denote $|\mathcal{Q}|$ and $|\mathcal{T}|$, respectively, as the size of the set of states $\mathcal{Q}$ and the set of transition relations $\mathcal{T}$. The computational burden of the algorithm in Paige and Tarjan (1987) to solve the RCP problem is $\mathcal{O}\left(|\mathcal{T}| \cdot \log(|\mathcal{Q}|)\right)$.

Example 6.4 Consider a labeled transition system $S = (\mathcal{Q}, \mathcal{A}, \mathcal{T}, q(0))$ shown in Fig. 6.1 with

$$\mathcal{Q} = \{0, 1, 2, 3, 4, 5, 6\}, \quad \mathcal{A} = \{a, b\},$$

Transitions for action a :

$$T_a[0] = \{3\}, \ T_a[1] = \{4\}, \ T_a[2] = \{4\}, \ T_a[3] = \{5\}, T_a[4] = \{6\}.$$

Transitions for action b :

$$T_b[0] = \{5\}, \ T_b[1] = \{6\}, \ T_b[2] = \{6\}, \ T_b[5] = \{6\}, T_b[6] = \{5\}.$$

The initial partition ρ is $\rho = \{B_1\}$ with $B_1 = \mathcal{Q} = \{0, 1, 2, 3, 4, 5, 6\}$. Applying the algorithm in Paige and Tarjan (1987) to solve the RCP problem, the coarsest partition ρ^* is $\rho^* = \{B_1^*, B_2^*, B_3^*\}$ with

$$B_1^* = \{0, 1, 2\}, \ B_2^* = \{3, 4\}, \ B_3^* = \{5, 6\}.$$

Next, an efficient approach for observability analysis will be proposed. As a BCN can be described by (2.32)–(2.33), the BCN is indeed a transition system with the set of states $\mathcal{Q} = \{\delta_{2^n}^1, \delta_{2^n}^2, \cdots, \delta_{2^n}^{2^n}\}$ and the set of actions $\mathcal{A} = \{\delta_{2^m}^1, \delta_{2^m}^2, \cdots, \delta_{2^m}^{2^m}\}$. The transition relation $T_u(x)$ is the set of successor states x' of the state x, i.e.

$$T_u(x) = \{x' | x' = L_{eq} \ltimes x \ltimes u\}. \tag{6.7}$$

Assume that $\rho = \{B_1, B_2, \cdots, B_{2^p}\}$ with B_i defined as

$$B_i = \{\delta_{2^n}^j | j \in \mathbb{Z}_+, Col_j(H) = \delta_{2^p}^i, 1 \leq j \leq 2^n\} \tag{6.8}$$

is an initial partition of the state space Δ_{2^n}. The states in the same block B_i generate the same output $\delta_{2^p}^i$. If there is no state that can generate the output $y = \delta_{2^p}^i$, $i \in \{1, 2, \cdots, 2^p\}$, then the block B_i is an empty set. In this case, the empty set B_i is removed from the partition ρ. Without loss of generality, suppose that all the blocks B_i in the partition ρ are not empty sets. Denote the coarsest refinement of the partition ρ (i.e. the coarsest partition refining the partition ρ according to the transition relation (6.7)) as

$$\rho^* = \{B_1^*, B_2^*, \cdots, B_{|\rho^*|}^*\}. \tag{6.9}$$

Then the following result can be obtained.

Theorem 6.5 *A BCN (2.32)–(2.33) is observable if and only if any block B_i^*, $i = 1, 2, \cdots, |\rho^*|$ in the coarsest partition ρ^* contains only one state, i.e. there are 2^n blocks in the coarsest partition ρ^*.*

Proof . (Necessity) Suppose, by contradiction, that some blocks B_i^* contain more than one state. Recall that the blocks $B_1, B_2, \cdots, B_{2^p}$ are initialized according to (6.8). Based on this, it can be concluded that the states in the same refinement of block $B_i^*, i = 1, 2, \cdots, |\rho^*|$ generate the same output. According to Definition 6.2, as the partition ρ^* is a refinement of the partition ρ based on the algorithm proposed in Paige and Tarjan (1987), the partition ρ^* is compatible with the transitions (6.7). For any two states $\bar{x}(0)$ and $x(0)$ in the block B_i^*, giving any input sequence $\{u(0), u(1), \cdots\}$, the successor states are also in the same block B_j^*. This implies that, starting with initial states $\bar{x}(0)$ and $x(0)$, the same input and output trajectories are generated which contradicts Definition 6.1.

(Sufficiency) Suppose, by contradiction, that the BCN is unobservable. According to Definition 6.1, there are at least two states $\bar{x}$ and x so that given any input sequence $\{u(0), u(1), \cdots\}$, the corresponding output sequences are $\{\bar{y}(0), \bar{y}(1), \cdots\} = \{y(0), y(1), \cdots\}$. As each block in the partition ρ^* contains only one state, $\bar{x}$ and x belong, respectively, to the blocks $B_{i_1}^*$ and $B_{j_1}^*$. Since the blocks in ρ are initialized as (6.8), one block $B_i, i \in \{1, 2, \cdots, 2^p\}$ can be split into B_{i_1} and B_{i_2} if and only if there is at least one input sequence so that starting with the states in the blocks B_{i_1} and B_{i_2} the corresponding output trajectories are different. As $\bar{x}$ and x belong to different blocks $B_{i_1}^*$ and $B_{j_1}^*$ in the partition ρ^*, the corresponding generated input and output trajectories are different. This contradicts the assumption. □

Note that if there is a block $B_i^*, i = 1, 2, \cdots, |\rho^*|$ in the coarsest partition ρ^* containing more than one state, then these states are indistinguishable and the BCN is unobservable. This provides a possibility to design a reduced-order observer for the fault detection introduced later.

Observability analysis problems for BCNs are in general $\mathcal{NP}$-hard (Laschov et al., 2013). If the approach proposed in this chapter is applied to observability analysis, then the computational burden is $\mathcal{O}(n \cdot 2^{n+m})$. Until now the existing approaches for observability analysis of BCNs requires $\mathcal{O}(2^{2n+m})$. As $n \cdot 2^{n+m} < 2^{2n+m}$, the approach proposed in this chapter works much more efficiently.

6.1.2 Passive Fault Detection

For fault detection, a state observer will be used to monitor systems. Two states of a system are indistinguishable, if beginning with these two states the input and output trajectories of the system coincide at any time instant. If a BCN is not observable, then there are some indistinguishable states (Fornasini and Valcher, 2013). For fault detection it is not necessary to exactly estimate all internal states. Therefore, for the unobservable BCN the indistinguishable states can be grouped to reduce computational complexity of fault diagnosis. Based on this, a reduced-order observer can be designed. In this section, to judge whether a fault can be detected by the approach introduced later, fault detectability in BCNs will be addressed at first. Then, an approach to design a residual generator based on reduced-order observer and the corresponding decision logic for fault detection of BCNs will be introduced.

Passive fault detectability

Active fault detection of BCNs has the purpose that considering a fault candidate, corresponding control action should be taken. Different from that, a passive fault detection approach determines fault occurrence by observing the input and output trajectory without interfering the normal system dynamic. In this part, condition will be studied, under which a fault occurrence can be detected. To this goal, the definition of passive fault detectability of BCNs is given as follows.

Definition 6.6 (Passive fault detectability) A fault is said to be passively detectable, if for each initial state there is an input input sequence $\{u(0), u(1), \cdots, u(T)\}$ the output trajectories $\{y(0), y(1), \cdots, y(T)\}$ and $\{y_F(0), y_F(1), \cdots, y_F(T)\}$ generated, respectively, by the non-faulty BCN and the faulty BCN differ at some time instant $t \in [0, T]$, no matter what the initial state is.

Next, for the passive fault detectability analysis, a RCP problem will be solved. By looking at Definition 6.6, it can be recognized that the passive fault detectability analysis can be reformulated as finding initial states, respectively, in the non-faulty and faulty system such that non-faulty and faulty systems generate the same input and output trajectories. Hence, similar to Theorem 6.5, one can initialize the partition ρ as (6.8) and solve the RCP problem by using the algorithm proposed in Paige and Tarjan (1987).

However, a non-faulty or faulty BCN may contain indistinguishable states. At first, the non-faulty and faulty models should be reduced, respectively, so that the reduced non-faulty and faulty BCNs are observable. In order to do non-faulty model reduction, initialize the partition of the state space Δ_{2^n} as $\rho = \{B_1, B_2, \cdots, B_{2^p}\}$ with the block B_i specified as (6.8). By applying the algorithm introduced in Paige and Tarjan (1987) and according to the transition relation (6.7), the coarsest refinement of the partition ρ, is found, which is denoted by

$$\rho^* = \{B_1^*, B_2^*, \cdots, B_{|\rho^*|}^*\}. \tag{6.10}$$

Based on the coarsest partition ρ^* in (6.10), a Boolean matrix $\bar{T}$ of dimensions $|\rho^*| \times 2^n$ can be constructed with the (i, j)-th entry calculated as

$$\bar{T}_{(i,j)} = \begin{cases} 1, \text{ if } \delta_{2^n}^q \in B_i^*, \\ 0, \text{ otherwise.} \end{cases} \tag{6.11}$$

where $i = 1, 2, \cdots, |\rho^*|$ and $q = 1, 2, \cdots, 2^n$. The matrix $\bar{T}$ associates the states in the set B_i^* with the state $z = \delta_{|\rho^*|}^i$, $i = 1, 2, \cdots, |\rho^*|$, i.e. $z(t) = \bar{T}x(t)$. Then, a reduced non-faulty model can be constructed as

$$z(t+1) = \bar{T}L_{eq}\bar{T}^{\mathrm{T}}z(t)u(t), \tag{6.12}$$

$$y(t) = H\bar{T}^{\mathrm{T}}z(t) \tag{6.13}$$

where $z(t)$ is state of the reduced non-faulty system. In the same way, a matrix $\bar{T}_F$ can be constructed for the fault model (6.1)–(6.2). The reduced faulty model is

$$z_F(t+1) = \bar{T}_F L_F \bar{T}_F^{\mathrm{T}} z_F(t) u(t), \tag{6.14}$$

$$y(t) = H_F \bar{T}_F^{\mathrm{T}} z_F(t) \tag{6.15}$$

where $z_f(t)$ is state of the reduced faulty system.

In the second step, let a new state $\tilde{x}$ be defined as a collection of the states z and z_F and take the following form

$$\tilde{x}(t) = \left[z^{\mathrm{T}}(t) \ \ z_F^{\mathrm{T}}(t) \right]^{\mathrm{T}}. \tag{6.16}$$

The dynamics of the state $\tilde{x}(t)$ can be described by

$$\tilde{x}(t+1) = \tilde{L}\tilde{x}(t)u(t), \tag{6.17}$$

$$y(t) = \tilde{H}\tilde{x}(t) \tag{6.18}$$

with the logical matrices $\tilde{L}$ and $\tilde{H}$ given as

$$\tilde{L} = \begin{bmatrix} \bar{T} L_{eq} \bar{T}^{\mathrm{T}} & \mathbf{0} \\ \mathbf{0} & \bar{T}_F L_F \bar{T}_F^{\mathrm{T}} \end{bmatrix}, \quad \tilde{H} = \left[H \bar{T}^{\mathrm{T}} \quad H_F \bar{T}_F^{\mathrm{T}} \right].$$

According to (2.1), (6.17) can be rewritten as

$$\tilde{x}(t+1) = \begin{bmatrix} \bar{T} L_{eq} \bar{T}^{\mathrm{T}} z(t) u(t) \\ \bar{T}_F L_F \bar{T}_F^{\mathrm{T}} z_F(t) u(t) \end{bmatrix}$$

which shows that if let z be a zero vector, then the BCN (6.17)–(6.18) describes the dynamics of the faulty system (6.1)–(6.2). The problem to check the passive fault detectability can be reformulated as finding the indistinguishable states of the BCN (6.17)–(6.18). Denote $\tilde{N}$ as the dimension of the vector $\tilde{x}$. Initialize the partition as $\tilde{\rho} = \{\tilde{B}_1, \tilde{B}_2, \cdots, \tilde{B}_{\tilde{N}}\}$ with $\tilde{B}_i$ defined as

$$\tilde{B}_i = \{\delta_{\tilde{N}}^j | j \in \mathbb{Z}_+, Col_j(\tilde{H}) = \delta_{2^p}^i, 1 \leq j \leq \tilde{N}\}. \tag{6.19}$$

Solve the RCP problem by using the algorithm proposed in Paige and Tarjan (1987). Denote the coarsest refinement of the partition $\tilde{\rho}^*$ as $\tilde{\rho}^* = \{\tilde{B}_1^*, \tilde{B}_2^*, \cdots, \tilde{B}_{|\tilde{\rho}^*|}^*\}$. Based on the partition $\tilde{\rho}^*$, one can draw the following conclusion.

Theorem 6.7 *A fault is passively detectable, if and only if any block* $\tilde{B}_i^*$, $i = 1, 2, \cdots, |\tilde{\rho}^*|$ *in the partition* $\tilde{\rho}^*$ *contains only one state in the set* $\Delta_{\tilde{N}}$.

Proof . (Sufficiency) Assume that all the blocks $\tilde{B}_i^*$, $i = 1, 2, \cdots, |\tilde{\rho}^*|$ in the partition $\tilde{\rho}^*$ contain only one state in the set $\Delta_{\tilde{N}}$. According to Theorem 6.5 the BCN (6.17)–(6.18) is observable. That means the non-faulty and faulty BCN cannot generate the same input and output trajectory, which corresponds to the abnormal behavior of the system. Based on this, the fault can be detected.

(Necessity) Suppose, by contradiction, that the fault is not passively detectable and all the blocks $\tilde{B}_i^*$ contain one state. According to Definition 6.6, there exists an input sequence $\{u(0), u(1), \cdot, u(T)\}$, such that the output trajectories $\{y(0), y(1), \cdots, y(T)\}$ and $\{y_F(0), y_F(1), \cdots, y_F(T)\}$ generated, respectively, by the non-fault BCN and the faulty BCN are the same. According to Theorem 6.5 the BCN (6.17)–(6.18) is unobservable. That means there is at least a block $\tilde{B}_i^*$ that contains more than one state. This contradicts the assumption. □

Remark 6.8 Theorem 6.7 gives a necessary and sufficient condition for passive fault detectability. In the literature, Fornasini and Valcher (2015a) has proposed an approach for passive fault detectability analysis. The key is to construct a non-faulty-faulty (NF-F) directed graph containing 2^{2n} nodes to represent all possible pairs of states of non-faulty and faulty BCN. Starting with any pair of states, the path will eventually enter the set of pairs of states that generate different outputs. In graph theory, one efficient way is to apply the Tarjan's algorithm (Tarjan, 1972) to find the strongly connected component in the NF-F graph. The corresponding computational burden is $\mathcal{O}(2^{2n+m})$. In comparison, for non-faulty and faulty model reduction, one needs $\mathcal{O}(n \cdot 2^{n+m})$ computational effort. To get the coarsest partition $\tilde{\rho}^*$ the computational cost is $\mathcal{O}(n \cdot 2^{n+m})$. Hence, the approach proposed in this paper runs in time $\mathcal{O}(n \cdot 2^{n+m})$, which is more efficient than the method given in Fornasini and Valcher (2015a).

Reduced-order observer based residual generator

In this section, an approach to design a residual generator based on reduced-order observer will be proposed. If a BCN is unobservable and a fault is detectable, then the approach proposed here works more efficiently than the residual generator based on the Luenberger-like observer. The basic idea is to get rid of the indistinguishable states.

Observer-based fault detection of BCNs was firstly studied in Fornasini and Valcher (2015a). In addition, in the same way as pointed out in Fornasini and Valcher (2015a), it is assumed that a BCN cannot recover itself autonomously after a fault occurs. A residual generator based on the Luenberger-like observer takes the following form:

$$\begin{aligned}
\hat{x}_s(0) &= H^{\mathrm{T}} y(0), \\
\hat{x}_s(t) &= \Phi_{2^n}^{\mathrm{T}} (I_{2^n} \otimes H^{\mathrm{T}}) L_{eq} \hat{x}_s(t-1) u(t-1) y(t), \\
r(t) &= 1 - \mathbf{1}_{2^n}^{\mathrm{T}} \hat{x}_s(t)
\end{aligned} \tag{6.20}$$

where $\hat{x}_s(t)$ is state estimate and $r(t) \in \mathbb{Z}$ is the residual signal. As Boolean product (2.36) is applied, the state estimate $\hat{x}_s(t) \in \mathcal{D}^{2^n} \setminus \{\mathbf{0}_{2^n}\}$ provided by the Luenberger-like observer is a Boolean vector. Accordingly, the residual signal $r(t)$ holds $r(t) < 1$ for the fault-free case. If a fault that influences system dynamic, has happened, then the vector $\hat{x}_s(t)$ will become a zero vector, i.e. $r(t) = 1$. Hence, the decision logic for the residual generator (6.20) is set to

$$\begin{cases} r(t) = 1 & \Rightarrow \text{A fault has occurred,} \\ r(t) < 1 & \Rightarrow \text{The BCN is fault-free.} \end{cases} \tag{6.21}$$

However, as a full-order observer is used, it requires high computational effort for high-dimensional systems. Therefore, an approach is introduced to design a residual generator based on a reduced-order observer to reduce the computational complexity for fault detection.

Assume that the matrix $\bar{T}$ has been constructed according to (6.11) for building a reduced non-faulty BCN (6.12)–(6.13). Based on the matrix $\bar{T}$, a residual generator can be constructed based on a reduced-order observer as

$$\begin{aligned}
\hat{z}_s(0) &= \bar{T} H^{\mathrm{T}} y(0), \\
\hat{z}_s(t) &= \bar{T} \Phi_{2^n}^{\mathrm{T}} (I_{2^n} \otimes H^{\mathrm{T}}) L_{eq} \bar{T}^{\mathrm{T}} \hat{z}_s(t-1) u(t-1) y(t), \\
r_z(t) &= 1 - \mathbf{1}_{|\rho^*|}^{\mathrm{T}} \hat{z}_s(t)
\end{aligned} \tag{6.22}$$

where $\hat{z}_s$ is state estimate of the state z of the reduced non-faulty BCN (6.12)–(6.13) and $r_z(t)$ is the residual signal. At time $t = 0$, $\hat{z}_s(t)$ is initialized as $\bar{T} H^{\mathrm{T}} y(0)$. If the i-th entry in $\bar{T} H^{\mathrm{T}} y(0)$ is nonzero, then $\rho^i_{|\bar{\rho}^*|}$ is a state candidate. The set of state candidates will be updated based on the new available information of input and output. If $r_z(t) = 1$, then it can be concluded that a fault has occurred, while $r_z(t) < 1$ shows that the BCN is fault-free. After getting the residual signal rz(t) provided by the residual generator based on reduced-order observer (6.22), we can draw the following conclusion.

Theorem 6.9 *If a detectable fault has occurred, then the residual signals $r(t)$ and $r_z(t)$ produced, respectively, by the residual generator based on the Luenberger-like observer* (6.20) *and the reduced-order observer* (6.22) *converge to the value* 1 *at the same time.*

Proof . The theorem will be proven by induction. It will be shown that $\hat{z}_s(t) = \bar{T}\hat{x}_s(t), t = 0, 1, \cdots$. Let's start with $t = 0$. After comparing (6.22) with (6.20), it can be directly got that at time $t = 0$ one has $\hat{z}_s(0) = \bar{T}\hat{x}_s(0)$.

Next, assume that $\hat{z}_s(t-1) = \bar{T}\hat{x}_s(t-1)$ holds. (6.22) can be equivalently written as

$$\hat{z}_s(t) = \bar{T}\Phi_{2^n}^{\mathrm{T}}(I_{2^n} \otimes H^{\mathrm{T}})L_{eq}\bar{T}^{\mathrm{T}}\bar{T}\hat{x}_s(t-1)u(t-1)y(t). \tag{6.23}$$

According to Lemma 2.10 and Proposition 2.5, (6.23) can be rewritten as

$$\hat{z}_s(t)=\bar{T}\left(\left(L_{eq}\bar{T}^{\mathrm{T}}\bar{T}\hat{x}_s(t-1)u(t-1)\right) \odot \left(H^{\mathrm{T}}y(t)\right)\right). \tag{6.24}$$

For the sake of simplicity, let $\hat{x}_s(t-1) \in B_i^*$ be the state estimate provided by the Luenberger-like observer and its successor state belongs to the block B_k^* (i.e. $L_{eq}\hat{x}_s(t-1)u(t-1) \in B_k^*$). Then $\bar{T}^{\mathrm{T}}\bar{T}\hat{x}_s(t-1)$ represents all the states of the block B_i^*. Recall that all the states in the same block of the partition ρ^* generate the same output. If $HL_{eq}\hat{x}_s(t-1)u(t-1) = y(t)$, then according to Boolean product (2.36) one has $\hat{z}_s(t) = \bar{T}L_{eq}\bar{T}^{\mathrm{T}}\bar{T}\hat{x}_s(t-1)u(t-1) = \bar{T}\hat{x}_s(t)$. Therefore, $\hat{z}_s(t) = \bar{T}\hat{x}_s(t)$ is obtained. It follows from the inductive step that

$$\hat{z}_s(t) = \bar{T}\hat{x}_s(t), t = 0, 1, \cdots. \tag{6.25}$$

If the state estimate $\hat{x}_s(t)$ becomes a zero vector at time t, then it can be concluded that a fault has happened not later than time t. According to (6.25), $\hat{z}_s(t)$ is a zero vector at time t. Accordingly, the residual signal $r_z(t)$ becomes one. □

Remark 6.10 The fault detection approach based on the reduced-order observer (6.22) requires online computation of complexity $\mathcal{O}(|\rho^*|^2 \cdot 2^{m+p})$, while the computational burden of the Luenberger-like observer based approach is $\mathcal{O}(2^{2n} \cdot 2^{m+p})$. If a BCN (2.32)–(2.33) is not observable, i.e. $|\rho^*| < 2^n$, then it is meaningful to construct a residual generator based on the reduced-order observer (6.22). Compared with the case of using the Luenberger-like observer (6.20), applying the reduced-order observer to fault detection can reduce the computational complexity by the factor $(2^n/|\rho^*|)^2$.

Example 6.11 In order to illustrate the result given in this section, consider the following BCN

$$\begin{cases} X_1(t+1) = U(t) \wedge \neg X_6(t), \\ X_2(t+1) = \neg X_1(t), \\ X_3(t+1) = \neg X_1(t) \wedge (X_5(t) \vee X_3(t)), \\ X_4(t+1) = X_1(t) \wedge \neg X_6(t), \\ X_5(t+1) = X_4(t) \vee \neg X_3(t), \\ X_6(t+1) = X_5(t) \wedge (\neg X_6(t) \vee \neg X_2(t)), \end{cases} \tag{6.26}$$

which is a Boolean model of the oxidative stress response pathways derived in Sridharan et al. (2012). In the model, U is the oxidative concentration. X_1 is a biochemical entity called reactive oxidative species. X_2 represents a transcription regulator protein. X_3 is a Kelch-like ECH-associated protein. X_4 is short for PKC which stands for Protein kinase C. X_5 represents the nuclear factor erythroid 2–related factor 2 while X_6 denotes an antioxidant response element. Assume that the state variables X_1, X_2 and X_6 can be measured, i.e.

$$Y_1(t) = X_1(t), Y_2(t) = X_2(t), Y_3(t) = X_6(t). \tag{6.27}$$

Using STP, the BCN (6.26)–(6.27) can be converted into the equivalent algebraic form (2.32)–(2.33) with $x(t) = \ltimes_{i=1}^{6} x_i(t) \in \Delta_{64}$, $u(t) \in \Delta_2$, $y(t) = \ltimes_{i=1}^{3} y_i(t)$ and the logical matrices $L_{eq} \in \mathcal{L}_{64\times 128}$ and $H \in \mathcal{L}_{8\times 64}$.

At first, the observability of the BCN (6.26)–(6.27) will be checked. For this purpose, let the partition ρ be initialized as (6.8), i.e. $\rho = \{B_1, B_2, B_3, B_4, B_5, B_6, B_7, B_8\}$ with

$$\begin{aligned}
B_1 =& \{\delta_{64}^{1}, \delta_{64}^{3}, \delta_{64}^{5}, \delta_{64}^{7}, \delta_{64}^{9}, \delta_{64}^{11}, \delta_{64}^{13}, \delta_{64}^{15}\}, \\
B_2 =& \{\delta_{64}^{2}, \delta_{64}^{4}, \delta_{64}^{6}, \delta_{64}^{8}, \delta_{64}^{10}, \delta_{64}^{12}, \delta_{64}^{14}, \delta_{64}^{16}\}, \\
B_3 =& \{\delta_{64}^{17}, \delta_{64}^{19}, \delta_{64}^{21}, \delta_{64}^{23}, \delta_{64}^{25}, \delta_{64}^{27}, \delta_{64}^{29}, \delta_{64}^{31}\}, \\
B_4 =& \{\delta_{64}^{18}, \delta_{64}^{20}, \delta_{64}^{22}, \delta_{64}^{24}, \delta_{64}^{26}, \delta_{64}^{28}, \delta_{64}^{30}, \delta_{64}^{32}\}, \\
B_5 =& \{\delta_{64}^{33}, \delta_{64}^{35}, \delta_{64}^{37}, \delta_{64}^{39}, \delta_{64}^{41}, \delta_{64}^{43}, \delta_{64}^{45}, \delta_{64}^{47}\}, \\
B_6 =& \{\delta_{64}^{34}, \delta_{64}^{36}, \delta_{64}^{38}, \delta_{64}^{40}, \delta_{64}^{42}, \delta_{64}^{44}, \delta_{64}^{46}, \delta_{64}^{48}\}, \\
B_7 =& \{\delta_{64}^{49}, \delta_{64}^{51}, \delta_{64}^{53}, \delta_{64}^{55}, \delta_{64}^{57}, \delta_{64}^{59}, \delta_{64}^{61}, \delta_{64}^{63}\}, \\
B_8 =& \{\delta_{64}^{50}, \delta_{64}^{52}, \delta_{64}^{54}, \delta_{64}^{56}, \delta_{64}^{58}, \delta_{64}^{60}, \delta_{64}^{62}, \delta_{64}^{64}\}.
\end{aligned}$$

By applying the algorithm proposed in Paige and Tarjan (1987) to solve the RCP problem, the coarsest partition ρ^* is got and has 27 blocks, i.e. $\rho^* = \{B_1^*, B_2^*, \cdots, B_{27}^*\}$. As $|\rho^*| = 27 < 64$, there exist indistinguishable states in the BCN (6.26)–(6.27). The BCN is unobservable. Hence, a fault detector based on reduced-order observer can be constructed. Let the matrix $\bar{T}$ of dimensions 27×64 be constructed according to (6.11). There is

$$\begin{array}{rl} \bar{T} = \delta_{27}[1 & 2\ \ 1\ \ 11\ 24\ 17\ 24\ 19\ \ 1\ \ 2\ \ 1\ \ 11\ \ 1 \\ & 2\ \ 1\ \ 11\ \ 3\ \ 4\ \ 14\ 12\ 23\ 18\ 25\ 20\ \ 3\ \ 4 \\ & 14\ 12\ \ 3\ \ 4\ \ 14\ 12\ \ 5\ \ 6\ \ 5\ \ 26\ 21\ \ 6\ \ 21 \\ & 15\ \ 5\ \ 6\ \ 5\ \ 9\ \ 5\ \ 6\ \ 5\ \ 9\ \ 7\ \ 8\ \ 13\ \ 27 \\ & 7\ \ 8\ \ 22\ 16\ \ 7\ \ 8\ \ 13\ 10\ \ 7\ \ 8\ \ 13\ \ 10]. \end{array}$$

After that, a fault detector based on the reduced-order observer (6.22) is created. In order to show the performance of the fault detector, assume that at time $t = 3$ a fault occurs and a measured input and output trajectory is given in Table 6.1.

Table 6.1 Input and output trajectory (vector form)

t	0	1	2	3	4	5	6	7	8	9	10	11	12
u	δ_2^1	δ_2^1	δ_2^1	δ_2^2	δ_2^1	δ_2^2	δ_2^2	δ_2^1	δ_2^2	δ_2^2	δ_2^2	δ_2^1	
y	δ_8^4	δ_8^3	δ_8^7	δ_8^5	δ_8^6	δ_8^2	δ_8^7	δ_8^5	δ_8^6	δ_8^5	δ_8^6	δ_8^5	δ_8^6

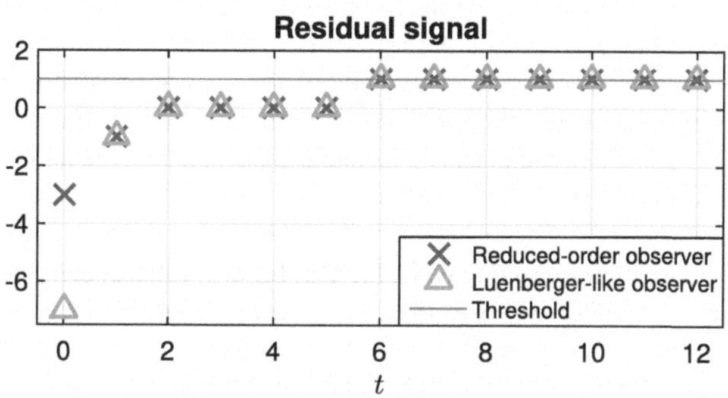

Figure 6.2 Residual signals provided by the residual generator based on observers (r: Luenberger-like observer based residual generator (6.20), r_z: reduced-order observer based residual generator (6.22)).

The residual signals generated, respectively, by the fault detector based on the Luenberger-like observer (6.20) and the reduced-order observer (6.22) are shown in Fig. 6.2. It can be seen that before time $t = 7$ the residual signals provided by the both fault detectors are smaller than the threshold. But after that, both fault detectors generate residual signals of value 1. According to the decision logic (6.21), fault is detected by the fault detectors based on the Luenberger-like observer (6.20) and the reduced-order observer (6.22) at the same time, which verifies the results given in Theorem 6.9. The system matrix of the reduced-order observer has dimension 27×432. In comparison, by the Luenberger-like observer a matrix of dimension 64×1024 is needed. Therefore, the fault detector based on the reduced-order observer (6.22) requires lower computational effort.

6.1.3 Passive Fault Isolation

Sometimes, a system cannot be simply shut down due to abnormal system behavior. For fault-tolerant control, after a fault has been detected, it is required to determine the faulty component through fault isolation with the help of residuals. However, it is not always possible to distinguish different types of faults. Hence, fault isolability will be analyzed at first. If the faults are not always isolable, then this means some faults have dynamic overlapping. In this case, the approach proposed in this chapter to design residual generators can be applied to execute fault isolation efficiently.

Fault isolability

Two systems are said to have dynamic overlapping if the two systems can generate the same admissible input and output trajectories. If there is dynamic overlapping between two faulty BCNs, then these two faults cannot always be distinguished. But if applying an input sequence the two faults can generate different output sequences, then it means that the two faults can be isolated. Based on this, the faulty isolability is defined as follows.

Definition 6.12 Faults are said to be isolable, if for any initial state of faulty BCNs there is an input sequence such that the corresponding output sequences differ.

Without loss of generality, for fault isolability analysis we assume that a fault has been already detected and all faults are not completely identical. Let a new state x_f be defined as a collection of the states x_{F_j}, $j = 1, 2, \cdots, k$ of the faulty BCN (6.3)–(6.4) that takes the following form

$$x_F(t) = \left[x_{F_1}^{\mathrm{T}}(t)\ x_{F_2}^{\mathrm{T}}(t)\ \cdots\ x_{F_k}^{\mathrm{T}}(t)\right]^{\mathrm{T}}. \tag{6.28}$$

In the same way as the construction of the BCN (6.17)–(6.18), the dynamics of the state $x_F(t)$ can be described by

$$x_F(t+1) = L_F x_F(t) u(t), \tag{6.29}$$

$$y(t) = H_F x_F(t) \tag{6.30}$$

with the logical matrices L_f and H_f given as

$$L_F = \mathrm{diag}\{L_{F_1}, L_{F_2}, \cdots, L_{F_k}\},$$

$$H_F = \left[H_{F_1}\ H_{F_2}\ \cdots\ H_{F_k}\right].$$

As a matter of fact, the BCN (6.29)–(6.30) contains the dynamics of all the faulty BCNs (6.3)–(6.4). If two faults in $\{F_1, F_2, \cdots, F_k\}$ are not isolable, than according to Definition 6.12 the BCN (6.29)–(6.30) is indeed not observable. Hence, the problem to analyze fault isolability can be reformulated as observability analysis problem of the BCN (6.29)–(6.30), which can be converted into a RCP problem.

According to (6.28), the state space $\Delta_{k\cdot 2^n}$ contains all the states in the faulty BCNs for the faults $F_1, F_2, \cdots, F_k$. Initialize the partition of state space $\Delta_{k\cdot 2^n}$ as $\rho_f = \{B_{F,1}, B_{F,2}, \cdots, B_{F,2^p}\}$ with $B_{F,i}$ defined as

$$B_{F,i} = \{\delta_{k\cdot 2^n}^l \,|\, l \in \mathbb{Z}_+, Col_l(H_F) = \delta_{2^p}^i, 1 \le l \le k\cdot 2^n\},\ i = 1, 2, \cdots, 2^p. \tag{6.31}$$

From (6.31), it can be seen that all the sets $B_{F,1}, B_{F,2}, \cdots, B_{F,2^p}$ do not intersect with each other. According to the BCN (6.29)–(6.30) and the partition ρ_F initialized by (6.31), the RCP problem is solved by using the algorithm proposed in Paige and Tarjan (1987) to find the coarsest refinement of the partition ρ_F denoted as $\rho_F^* = \{B_{F,1}^*, B_{F,2}^*, \cdots, B_{F,|\rho_F^*|}^*\}$. Based on the coarsest refinement ρ_F^*, the following result can be achieved.

Theorem 6.13 *The faults $F_1, F_2, \cdots, F_k$ are isolable if and only if each block $B_{F,j}^*$, $j = 1, 2, \cdots, |\rho_f^*|$ in the relational coarsest partition ρ_F^* contains only the part of the states in one faulty BCN* (6.3)–(6.4).

Proof . This theorem can be proven in the same way as Theorem 6.5. Hence, the proof is omitted here. □

Theorem 6.13 gives the necessary and sufficient condition of fault isolability. If the condition given in Theorem 6.13 is not satisfied, then whether faults can be isolated, depends on the initial states and the input sequence.

Design of reduced-order fault isolation filters

If a part of the faults cannot be isolated from each other, then reduced-order fault isolation filters can be constructed, which work more efficiently than the bank of full-order observers. The basic idea is to group the faults into different sets according to their partly dynamic overlapping. The residual generators should be designed so that each residual signal is affected by the partly dynamic overlapping among the faults in one specific set.

Based on the partition ρ_F^*, a matrix $\bar{T}_F$ can be constructed in the same way as the matrix $\bar{T}$ given in (6.11). The logical matrices $L_{F,R}$ and $H_{F,R}$ of the new reduced complete model are

$$L_{F,R} = \bar{T}_F L_F \bar{T}_F^{\mathrm{T}} \text{ and } H_{F,R} = H_F \bar{T}_F^{\mathrm{T}}. \tag{6.32}$$

In graph theory, a directed graph can be denoted by $G = (\mathcal{V}, \mathcal{E})$ where $\mathcal{V}$ is a set of nodes and $\mathcal{E} \subseteq \mathcal{V} \times \mathcal{V}$ is a set of edges. As mentioned in Liang et al. (2017), the logical matrix $L_{F,R}$ can be associated with a directed graph G with a set of nodes $\mathcal{V} = \{1, 2, \cdots, |\rho_F^*|\}$. The edge (i, j) belongs to the set $\mathcal{E}$ if and only if $\left(L_{F,R}\right)_{(i,j)} = 1$. A weakly connected graph (component) refers to a undirected graph (subgraph) whose nodes are connected, no matter which direction the edge is Tarjan (1972). Note that node i represents the set of states $B_{F,i}^*$. If the nodes i and j do not belong to the same weakly connected component in the graph G, then it means that the states in the set $B_{F,i}^*$ can be neither steered to any state in the set $B_{F,j}^*$ nor reached from any state in the set $B_{F,j}^*$. That is to say each weakly connected component in the graph G corresponds to the dynamic of one independent subsystem and the number of weakly connected components in the graph G is equal to the number of independent subsystems. To find out the weakly connected components in the graph G, one can construct an undirected graph from a directed graph by removing the direction of all edges and then use the Tarjan's algorithm (Tarjan, 1972).

Each weakly connected component builds an independent subsystem. For each independent subsystem a reduced-order observer will be constructed. The key is to pick out the dynamic built by each weakly connected components. For the sake of clarity, assume that there are α weakly connected components. The i-th weakly connected component consists of τ_i nodes, which are associated with the sets $B^*_{F,i_1}, B^*_{F,i_2}, \cdots, B^*_{F,i_{\tau_i}}$. The i-th subsystem built according to the i-th weakly connected component can be represented by

$$x_{F,sub_i}(t+1) = L_{F,sub_i} x_{F,sub_i}(t) u(t), \tag{6.33}$$

$$y(t) = H_{F,sub_i} x_{F,sub_i}(t), i = 1, 2, \cdots, \alpha. \tag{6.34}$$

Let a matrix $\bar{T}_{sub_i}$ be defined as

$$\bar{T}_{sub_i} = \left[\delta_{|\rho^*_F|}[i_1 \ i_2 \ \cdots \ i_{\tau_i}]\right]^{\mathrm{T}}. \tag{6.35}$$

The logical matrices L_{F,sub_i} and H_{F,sub_i} for the i-th subsystem (6.33)–(6.34) are defined as

$$L_{F,sub_i} = \bar{T}_{sub_i} L_{F,R} \bar{T}^{\mathrm{T}}_{sub_i}, \ H_{F,sub_i} = H_{F,R} \bar{T}^{\mathrm{T}}_{sub_i}. \tag{6.36}$$

Then the residual generator for the i-th independent subsystem will be constructed as

$$\begin{aligned} \hat{x}_{F,sub_i}(0) &= H^{\mathrm{T}}_{F,sub_i} \ y(0), \\ \hat{x}_{F,sub_i}(t) &= \Phi^{\mathrm{T}}_{\tau_i}(I_{\tau_i} \otimes H^{\mathrm{T}}_{F,sub_i}) L_{F,sub_i} \hat{x}_{F,sub_i}(t-1) u(t-1) y(t), \\ r_{F,sub_i}(t) &= 1 - \mathbf{1}^{\mathrm{T}}_{\tau_i} \hat{x}_{F,sub_i}(t) \end{aligned} \tag{6.37}$$

where $\hat{x}_{F,sub_i}(t)$ is the state estimate for the i-th subsystem (6.33)–(6.34) and $r_{F,sub_i}(t) \in \mathbb{Z}$ is the residual signal, $i = 1, 2, \cdots, \alpha$.

For the sake of convenience, the procedure to construct residual generators to isolate different sets of faults is summarized in Algorithm 6.

After the residual generators in the form of (6.37) for all the independent subsystems generate residual signals, it is necessary to evaluate the influence of different faults on residual signals. According to the BCN (6.33)–(6.34) and the transition relation defined in (6.7), if the sets $B^*_{F,i_1}, B^*_{F,i_2}, \cdots, B^*_{F,i_{\tau_i}}$ in the i-th subsystem contain some states in the j-th faulty BCN (6.3)–(6.4), then the residual signal r_{F,sub_i} will be influenced by the fault F_j.

Therefore, it is found out which faults have influence on the residual signal r_{F,sub_i} by checking to which faulty BCNs (6.3)–(6.4) the states in the sets $B^*_{F,i_1}, B^*_{F,i_2}, \cdots, B^*_{F,i_{\tau_i}}$ belong. Based on the analysis of the α subsystems for the k faults, a logic table for fault isolation can be obtained.

Algorithm 6: Given k faulty models described by (6.3)–(6.4). Construct the residual generators for fault isolation.

---

1. Build the model (6.29)–(6.30).
2. Solve the RCP problem of the system (6.29)–(6.30).
3. Construct the matrix $\bar{T}_F$ according to (6.11).
4. Calculate the logical matrices $L_{F,R}$ and $H_{F,R}$ according to (6.32).
5. Create the graph G associated with $L_{F,R}$ and determine the weakly connected components in the graph G by applying Tarjan's algorithm (Tarjan, 1972).
6. Construct the matrix $\bar{T}_{sub_i}$ according to (6.35) for each weakly connected component.
7. Compute the logical matrices L_{F,sub_i} and H_{F,sub_i} based on (6.36).
8. Construct residual generators for each independent subsystem according to (6.37).

As a matter of fact, a residual signal r_{F,sub_i} can be sensitive to several faults. Denote $\mathcal{S}_i$ as the set of faults that the residual signal r_{F,sub_i} is sensitive to. Note that each fault belongs to one or more sets in $\mathcal{S}_1, \mathcal{S}_2, \cdots, \mathcal{S}_\alpha$. Each set $\mathcal{S}_1, \mathcal{S}_2, \cdots, \mathcal{S}_\alpha$ may contain more than one fault.

Recall that Boolean product (2.36) is applied. The vectors $\hat{x}_{F,sub_i}(t), t = 0, 1, \cdots$ are Boolean vectors and the residual signal $r_{F,sub_i}(t)$ will be always lower than or equal to 1. If a fault F_j has occurred and the sets $B^*_{F,i_1}, B^*_{F,i_2}, \cdots, B^*_{F,i_{\tau_i}}$ in the i-th subsystem do not contain any state in the j-th faulty BCN (6.3)–(6.4), then the vector $\hat{x}_{F,sub_i}(t)$ will become a zero vector. Accordingly, the residual signal $r_{F,sub_i}(t)$ will be 1. Based on this, one can apply the following decision logic to isolate the set of faults $\mathcal{S}_i$ from the other sets $\mathcal{S}_j, j \neq i$:

$$\begin{cases} r_{F,sub_i}(t) = 1 \Rightarrow & \text{All faults in } \mathcal{S}_i \text{ cannot} \\ & \text{have happened,} \\ r_{F,sub_i}(t) < 1 \Rightarrow & \text{At least one fault in the} \\ & \text{set } \mathcal{S}_i \text{ may have happened.} \end{cases} \tag{6.38}$$

The subsystems built by the weakly connected components have different dynamics. As different residual generator (6.37) is constructed based on the BCN (6.33)–(6.34) describing dynamics of different subsystems, any two residual signals r_{F,sub_i} and r_{F,sub_j} are sensitive to different system dynamics. Therefore, only one residual signal can be eventually lower than 1, even if a fault belongs to more than one set in $\mathcal{S}_1, \mathcal{S}_2, \cdots, \mathcal{S}_\alpha$.

Now the complexity of the online computation in the conventional scheme of a bank of full-order observers and the approach proposed here is analyzed. The computational burden of Luenberger-like observer (6.20) is $\mathcal{O}(2^{2n+m+p})$. Therefore, the conventional scheme of a bank of k Luenberger-like observers requires $\mathcal{O}(k \cdot 2^{2n+m+p})$ in total. Recall that the i-th BCN (6.33)–(6.34) has τ_i states. It can be obtained directly that $\tau_i \leq 2^n, i = 1, 2, \cdots, \alpha$. According to (6.37), $\Phi_{\tau_i}^{\mathrm{T}}(I_{\tau_i} \otimes H_{F,sub_i}^{\mathrm{T}})L_{F,sub_i}$ is a matrix of dimensions $\tau_i \times (\tau_i \cdot 2^{m+p})$. For each residual generator (6.37), one needs to multiply $\Phi_{\tau_i}^{\mathrm{T}}(I_{\tau_i} \otimes H_{F,sub_i}^{\mathrm{T}})L_{F,sub_i}$ with the vector $\hat{x}_{F,sub_i}(t-1)u(t-1)y(t)$. This requires $\mathcal{O}(\tau_i^2 \cdot 2^{m+p})$ computational effort. As a result, computational burden of the approach proposed in this paper is $\mathcal{O}((\tau_1^2 + \tau_2^2 + \cdots + \tau_\alpha^2) \cdot 2^{m+p})$. Recall that the faults $F_1, F_2, \cdots, F_k$ are not isolable. Without loss of generality, assume that the j-th fault F_j belongs to the sets $\mathcal{S}_{j_1}, \mathcal{S}_{j_2}, \cdots, \mathcal{S}_{j_{q_j}}$ and the $j_1, j_2, \cdots, j_{q_j}$-th subsystems have, respectively, $\tau_{j_1}, \tau_{j_2}, \cdots, \tau_{j_{q_j}}$ states. Note that the faulty BCN (6.29)–(6.30) may be unobservable. By solving the RCP problem the state space Δ_{2^n} of the faulty BCN for F_j is partitioned into q_j blocks. Therefore, one has

$$2^n \geq \tau_{j_1} + \tau_{j_2} + \cdots + \tau_{j_{q_j}}. \tag{6.39}$$

As $\tau_{j_1}, \tau_{j_2}, \cdots, \tau_{j_{q_j}}$ are positive, it is clear that

$$2^{2n} \geq \left(\tau_{j_1} + \tau_{j_2} + \cdots + \tau_{j_{q_j}}\right)^2 > \tau_{j_1}^2 + \tau_{j_2}^2 + \cdots + \tau_{j_{q_j}}^2. \tag{6.40}$$

Recall that each set $\mathcal{S}_1, \mathcal{S}_2, \cdots, \mathcal{S}_\alpha$ may contain more than one fault. For each τ_i, $i = 1, 2, \cdots, \alpha$ there is an index $j \in \{1, 2, \cdots, k\}$, such that $\tau_i \in \{\tau_{j_1}, \tau_{j_2}, \cdots, \tau_{j_{q_j}}\}$. Sum up (6.40) for the k faults together and one gets

$$k \cdot 2^{2n} > \sum_{j=1}^{k} \tau_{j_1}^2 + \tau_{j_2}^2 + \cdots + \tau_{j_{q_j}}^2 > \tau_1^2 + \tau_2^2 + \cdots + \tau_\alpha^2. \tag{6.41}$$

Therefore, the approach proposed in the paper needs much less online calculation than the conventional scheme of a bank of full-order observers.

It is important to note that in comparison to the conventional scheme of a bank of full-order observers, applying the approach proposed in this paper for fault isolation can reduce the computational effort independent of the number of subsystems under two circumstances. The first case is when one of the k faulty BCNs (6.3)–(6.4) is not observable. The reason is that one executes model reduction by getting rid of the indistinguishable states in the unobservable BCN for fault diagnosis. The second case is when the faulty BCNs (6.3)–(6.4) have dynamic overlapping in case of different faults. By solving the RCP problem, the dynamic overlapping among the faulty BCNs are found. Then, the whole dynamics are decomposed into several weakly connected components. By doing this, the dynamic overlapping is eliminated.

Remark 6.14 The α residual generators (6.37) for fault isolation are used after fault detection. If the residual signal r_{F,sub_i} generated by the residual generator (6.37) for the i-th subsystem is sensitive to all the faults, then one cannot use the residual signal r_{F,sub_i} to isolate any faults. Therefore, r_{F,sub_i} needs not be considered any more.

Example 6.15 Consider the BCN (6.26)–(6.27) given in Example 6.11. For fault isolation, suppose that there are four possible fault candidates. In case of the first fault F_1, the state variable X_4 is assumed to be stuck at the values 1 (i.e. the logical function f_4 is changed to $X_4(t) = 1$). As the second fault F_2, the state variable X_7 is tied to a logic 0, i.e. $X_7 = 0$. For the third fault F_3, the state variable X_4 is updated according to a new Boolean function $X_4(t+1) = X_1(t)$. As the fourth fault F_4, the Boolean function $X_5(t+1) = \neg X_3(t)$ is assigned to X_5.

In the first step, the fault isolability will be checked. To achieve this goal, the BCN (6.29)–(6.30) is constructed which has in total $4 \times 2^6 = 256$ states. By applying the algorithm proposed in Paige and Tarjan (1987) to solve the RCP problem, the 256 states are partitioned into 60 blocks in the coarsest partition ρ_F^*, i.e. $\rho_F^* = \{B_{F,1}^*, B_{F,2}^*, \cdots, B_{F,60}^*\}, |\rho_F^*| = 60$. As the blocks $B_{F,34}^*, B_{F,35}^*, \cdots, B_{F,58}^*$ contain the states in the 3-th and the 4-th faulty BCNs, according to Theorem 6.13 the faults F_1, F_2, F_3 and F_4 are not isolable.

In the next step, residual generators will be created. Construct the matrix $L_{F,R}$ according to (6.32). Let a directed graph G be built according to the logical matrix $L_{F,R}$. The direction of all edges in the graph G is removed and Tarjan's algorithm is applied to find the weakly connected components. As a result, the nodes associated with the sets in the partition ρ_F^* build 3 weakly connected components shown in Fig. 6.3, 6.4 and 6.5 (i.e. there are three subsystems; $\alpha = 3$). The first weakly connected component contains the nodes associated with the sets $B_{F,1}^*, B_{F,2}^*, \cdots, B_{F,25}^*$. The second weakly connected component is made up of the nodes associated with the sets $B_{F,26}^*, B_{F,27}^*, \cdots, B_{F,33}^*$, while in the third weakly connected component the nodes associated with the sets $B_{F,34}^*, B_{F,35}^*, \cdots, B_{F,60}^*$ are connected.

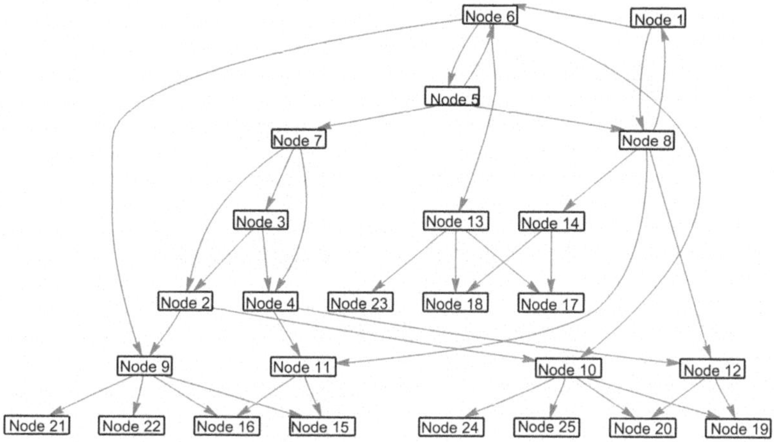

Figure 6.3 The first weakly connected component

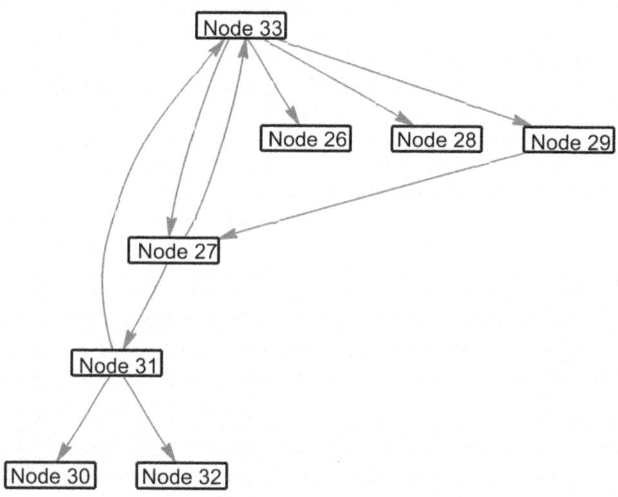

Figure 6.4 The second weakly connected component

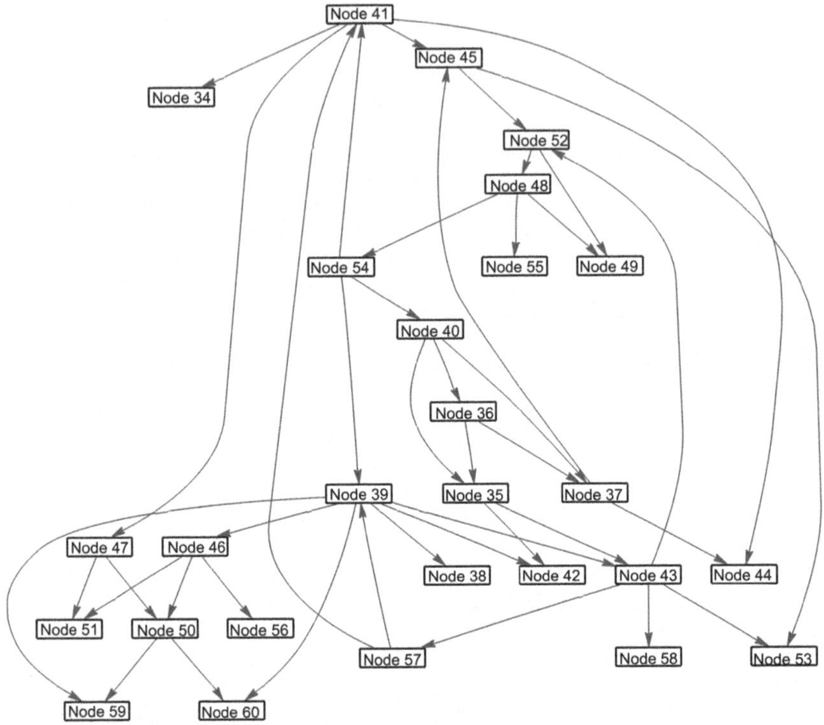

Figure 6.5 The third weakly connected component

Let three independent subsystems be built according to each weakly connected component. The dynamics of the subsystems can be described in the form of (6.33)–(6.34). The subsystem 1 has 25 states and the logical matrices L_{F,sub_1} and H_{F,sub_1} have, respectively, the dimensions 25×50 and 25×25. The second subsystem has 8 states and the logical matrices L_{F,sub_2} and H_{F,sub_2} have, respectively, the dimensions 8×16 and 8×8. The third subsystem has 27 states and the logical matrices L_{F,sub_3} and H_{F,sub_3} have, respectively, the dimensions 27×54 and 27×27.

By looking at the blocks $B^*_{F,1}, B^*_{F,2}, \cdots, B^*_{F,60}$, it is found out that the sets $B^*_{F,1}, B^*_{F,2}, \cdots, B^*_{F,25}$ contain only the states in the faulty BCN for the fault F_1. The states in the faulty BCN for the fault F_2 are only contained in the sets $B^*_{F,26}, B^*_{F,27}, \cdots, B^*_{F,33}$. The sets $B^*_{F,34}, B^*_{F,35}, \cdots, B^*_{F,60}$ are collection of the states in the faulty BCN for the faults F_3 and F_4. Therefore, the subsystems 1 and

2 describe, respectively, the dynamics of the faults F_1 and F_2, while the subsystem 3 represents the dynamics of the faults F_3 and F_4. That means, there are three sets of faults that can be isolated and they are

$$\{F_1\}, \{F_2\} \text{ and } \{F_3, F_4\}.$$

For fault isolation, three residual generators can be constructed based on (6.37) with the residual signal r_{F,sub_1} sensitive to the fault F_1, the residual signal r_{F,sub_2} sensitive to the fault F_2 and the residual signal r_{F,sub_3} sensitive to the faults F_3 and F_4. Therefore, $\mathcal{S}_1 = \{F_1\}$, $\mathcal{S}_2 = \{F_2\}$, $\mathcal{S}_3 = \{F_3, F_4\}$ and the logic table are shown in Table 6.2.

Table 6.2 Logic table for fault isolation

	F_1	F_2	F_3	F_4
$r_{F,sub1}$	X			
$r_{F,sub2}$		X		
$r_{F,sub3}$			X	X

As mentioned in Example 6.11, the fault F_1 occurs at time $t = 3$. After the fault has been detected at time $t = 6$, the residual generators are implemented in Matlab to evaluate the input and output trajectory $\{(u(t), y(t)), t = 6, 7, \cdots, 12\}$ given in Example 6.11 (see Table 6.1). The generated residual signals are shown in Fig. 6.6. At time $t = 6$, all residual signals r_{F,sub_1}, r_{F,sub_2} and r_{F,sub_3} are lower than 1. This means that all faults may have happened. At time $t = 7, 8$, the residual signal r_{F,sub_2} arrives at the value 1. Hence, the set $\mathcal{S}_2$ (i.e. fault F_2) cannot have happened. At time $t = 9$, the residual signal r_{F,sub_3} reaches the value 1 and only the residual signal r_{F,sub_1} is different from 1. Therefore, according to the decision logic (6.38) and the logic table shown in Table 6.2, it can be concluded that the faults in the set $\mathcal{S}_3$ (i.e. faults F_3 and F_4) cannot have happened. As a result, the first fault F_1 has happened.

To show running time of applying the reduced-order observer for fault isolation, the procedure is repeated in Matlab for 20 times using the CPU Intel Core $i5-7500@3.40GHz$. As a result, the mean and the standard deviation of the CPU time by using the reduced-order observer are, respectively, $0.0034s$ and $7.7868 \times 10^{-4}s$.

In comparison, the conventional fault isolation scheme would require four Luenberger-like observers, each of which is constructed for one faulty model. After evaluating the same input and output trajectory in Table 6.1, the four residual signals r_1, r_2, r_3 and r_4 are depicted in Fig. 6.7. Recall that the state estimate provided by the Luenberger-like observer is a Boolean vector. Therefore, the upper bound of the residual signal is 1. If the Luenberger-like observer evaluates an inadmissible input and output trajectory, then the state estimate will be a zero vector. Correspondingly, the residual signal is 1. Thus, one can use the following decision logic:

$$\begin{cases} r_i(t) = 1 \Rightarrow \text{The fault } F_i \text{ cannot happened,} \\ r_i(t) < 1 \Rightarrow \text{The fault } F_i \text{ may have happened.} \end{cases} \tag{6.42}$$

From Fig. 6.7, it can be recognized that at time $t = 9$ only the residual signal r_1 is less than 1. This means that the conventional scheme would also identify the fault F_1 as the fault causing the deviation of the output sequence from normal system dynamics. However, the conventional scheme requires to use four Luenberger-like observers and each of them has a matrix of dimensions 64×1024. Different from that, three reduced-order observers (6.37) have matrices, respectively, of dimensions 25×400, 8×128 and 27×432. Hence, the approach proposed in this chapter needs much less online calculation than the conventional scheme.

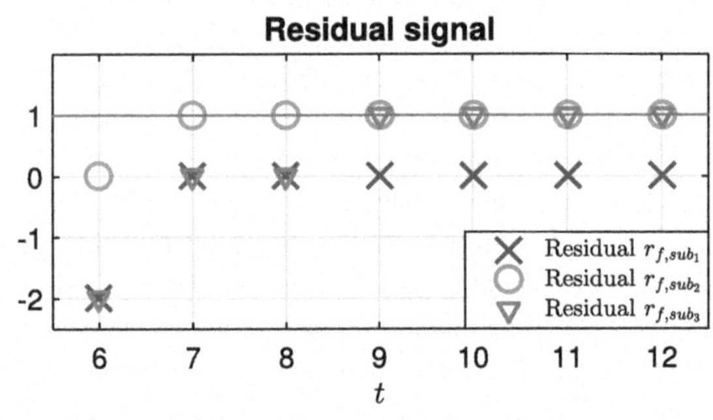

Figure 6.6 Residual signals generated by residual generators (6.37) for fault isolation.

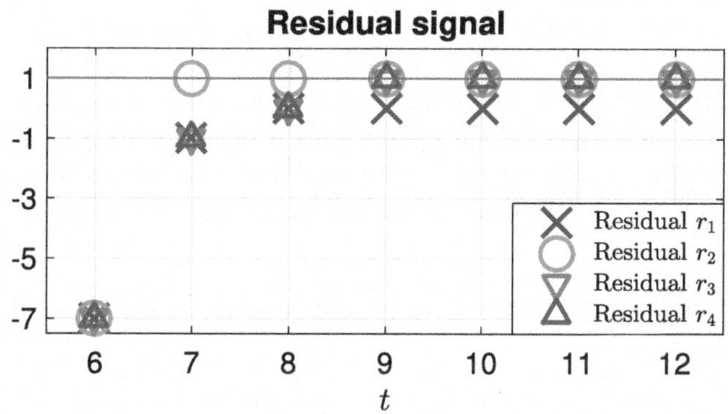

Figure 6.7 Residual signals generated by the bank of four Luenberger-like observers (6.20).

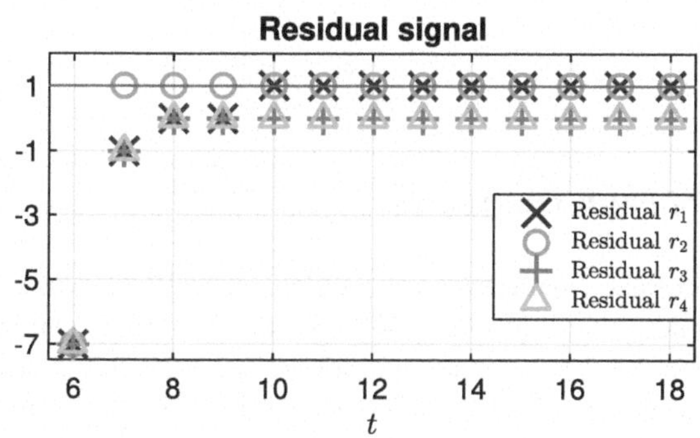

Figure 6.8 Residual signals delivered by the bank of four Luenberger-like observers (6.20) after evaluating the input and output trajectory given in Table 6.3.

As shown in Table 6.2, the proposed fault isolation scheme can not isolate the faults F_3 and F_4, because in the coarsest partition ρ_F^* the blocks $B_{F,34}^*$, $B_{F,35}^*$, $\cdots$, $B_{F,60}^*$ contain both the states in the faulty BCN for F_3 and the states in the faulty BCN for F_4. This means that the faulty models for the faults F_3 and F_4 have complete dynamic overlapping. It is worth noticing that the conventional scheme of four Luenberger-like observers can also not isolate the faults F_3 and F_4, though four observers have been used. Let us look at the following example where at time $t = 6$ the fault F_3 has occurred and the corresponding input and output trajectory $\{(u(t), y(t)), t = 0, 1, \cdots, 18\}$ is given in Table 6.3. The residual signals delivered by the bank of four Luenberger-like observers are shown in Fig. 6.8. From Fig. 6.8

it can be seen that after time $t = 10$ the residual signals r_1 and r_2 have reached the value 1, while the residual signals r_3 and r_4 are always less than 1. Hence, according to the residuals r_1, r_2, r_3 and r_4 in Fig. 6.8, it can be concluded that the faults F_1 and F_2 can not have happened, the fault that has happened is F_3 or F_4. But it also can not tell exactly whether F_3 or F_4 is the true fault. That means, F_3 and F_4 can also not be isolated with the bank of four Luenberger-like observers.

Table 6.3 Input and output trajectory (vector form) measured in the case that the fault F_3 occurs at time $t = 6$

t	0	1	2	3	4	5	6	7	8	9	10
u	δ_2^2	δ_2^2	δ_2^1	δ_2^1	δ_2^2	δ_2^1	δ_2^2	δ_2^2	δ_2^2	δ_2^1	δ_2^2
y	δ_8^2	δ_8^8	δ_8^5	δ_8^6	δ_8^2	δ_8^8	δ_8^2	δ_8^7	δ_8^5	δ_8^6	δ_8^2
t	11	12	13	14	15	16	17	18			
u	δ_2^1	δ_2^1	δ_2^2	δ_2^2	δ_2^2	δ_2^1	δ_2^1				
y	δ_8^8	δ_8^2	δ_8^3	δ_8^7	δ_8^5	δ_8^6	δ_8^2	δ_8^4			

6.2 Active Fault Diagnosis

In this section, the active fault diagnosis problem of the BCNs will be investigated. The active fault detection problem of BCNs studied is to design an input sequence so that a fault can be detected starting with any possible initial state. For active fault detection, a concept of active fault detectability of BCNs is introduced. After reformulating the active fault detection problem of BCNs as a dead-beat stabilization problem of autonomous switched linear discrete-time systems (ASLSs), a necessary and sufficient condition for active fault detectability analysis is given. After that, based on the result introduced in Fiacchini and Millérioux (2019), a computationally efficient approach is proposed to determine an input sequence of minimal length that achieves active fault detection of BCNs. Then an approach for active fault isolation of BCNs is proposed.

Note that the faulty BCN (6.1)–(6.2) is a more general faulty model than the one described in Fornasini and Valcher (2015a) where faults can only occur in state equation (6.1). In addition, as mentioned in Fornasini and Valcher (2015a), it is assumed that a BCN cannot recover itself autonomously after a fault occurs.

6.2.1 Active Fault Detectability Analysis

The concept of active fault detectability is introduced as follows.

Definition 6.16 (Active Fault Detectability) A fault F is said to be actively detectable, if there is an input sequence $\{u(0), u(1), \cdots, u(T)\}$ so that the output trajectories $\{y(0), y(1), \cdots, y(T+1)\}$ and $\{y_F(0), y_F(1), \cdots, y_F(T+1)\}$ generated, respectively, by the non-faulty BCN (2.32)–(2.33) and the faulty BCN (6.1)–(6.2) differ at some time instant t with $t \in [0, T+1]$, no matter what the initial state is.

Remark 6.17 It is worth mentioning that the actively detectable fault in Definition 6.16 is different from the detectable meaningful fault given in Fornasini and Valcher (2015a) which depends on the initial state.

In order to determine active fault detectability with respect to a given fault modeled by the BCN (6.1)–(6.2), an auxiliary state variable is defined as given in Fornasini and Valcher (2015b):

$$\tilde{x}(t) = x(t)x_F(t) \in \Delta_{2^{2n}}. \tag{6.43}$$

Applying Lemma 2.2 and the auxiliary state variable defined in (6.43) and recalling (2.32) and (6.1), there is

$$\begin{aligned}
\tilde{x}(t+1) &= x(t+1)x_F(t+1) \\
&= L_{eq}x(t)u(t)L_F x_F(t)u(t) \\
&= L_{eq}(I_{2^{m+n}} \otimes L_F)u(t)x(t)u(t)x_F(t) \\
&= L_{eq}(I_{2^{m+n}} \otimes L_F)u(t)W_{[2^m,2^n]}u(t)x(t)x_F(t) \\
&= L_{eq}(I_{2^{m+n}} \otimes L_F)(I_{2^m} \otimes W_{[2^m,2^n]})\Phi_{2^m}u(t)\tilde{x}(t).
\end{aligned} \tag{6.44}$$

For the sake of simplicity, let $\tilde{L} = L_{eq}(I_{2^{m+n}} \otimes L_F)(I_{2^m} \otimes W_{[2^m,2^n]})\Phi_{2^m}$. Then, according to (6.44), the augmented BCN can be written as given in Fornasini and Valcher (2015b):

$$\tilde{x}(t+1) = \tilde{L}u(t)\tilde{x}(t). \tag{6.45}$$

In thc next, a set of states $x^*(t)x_F^*(t) \in \Delta_{2^{2n}}$ which lead to $Hx^*(t) \neq H_F x_F^*(t)$ shall be determined. Given any two states $x(t) \in \Delta_{2^n}$ and $x_F(t) \in \Delta_{2^n}$. Since the outputs $y(t)$ and $y_F(t)$ belong to the set Δ_{2^p}, generally the result of the element-wise multiplication of the vectors $y(t) = Hx(t)$ and $y_F(t) = H_F x_F(t)$ (i.e. $y(t) \odot y_F(t) = Hx(t) \odot H_F x_F(t)$) belongs to the set $\Delta_{2^p} \cup \{\mathbf{0}_{2^p}\}$. If $Hx(t) \odot H_F x_F(t) = \mathbf{0}_{2^n}$, then the outputs $y(t)$ and $y_F(t)$ are different (i.e. $y(t) \neq y_F(t)$). Otherwise, if $Hx(t) \odot H_F x_F(t) \in \Delta_{2^p}$, then the outputs $y(t)$ and $y_F(t)$ are the same (i.e. $y(t) = y_F(t)$). Based on this fact, a signal $\tau(t) \in \mathcal{D}$ can be obtained as

$$\tau(t) = 1 - \mathbf{1}_{2^p}^{\mathrm{T}} \Big(Hx(t) \odot H_F x_F(t) \Big). \tag{6.46}$$

If $\tau(t) = 1$, then the outputs $y(t)$ and $y_F(t)$ do not coincide at time t. Otherwise the outputs $y(t)$ and $y_F(t)$ are equal. Applying Proposition 2.7 and Lemma 2.9, (6.46) can be expressed in the following algebraic form:

$$\begin{aligned} \tau(t) &= 1 - \mathbf{1}_{2^p}^{\mathrm{T}} \Phi_{2^p}^{\mathrm{T}} Hx(t) H_F x_F(t) \\ &= 1 - \mathbf{1}_{2^p}^{\mathrm{T}} \Phi_{2^p}^{\mathrm{T}} H(I_{2^n} \otimes H_F) x(t) x_F(t). \end{aligned} \tag{6.47}$$

Notice that $1 = \mathbf{1}_{2^{2n}}^{\mathrm{T}} x(t) x_f(t)$. Hence, (6.47) can be equivalently rewritten as

$$\tau(t) = \left(\mathbf{1}_{2^{2n}}^{\mathrm{T}} - \mathbf{1}_{2^p}^{\mathrm{T}} \Phi_{2^p}^{\mathrm{T}} H(I_{2^n} \otimes H_F) \right) x(t) x_F(t). \tag{6.48}$$

For the sake of simplicity, let the matrix $\tilde{H}$ be defined as

$$\tilde{H} = \mathbf{1}_{2^{2n}}^{\mathrm{T}} - \mathbf{1}_{2^p}^{\mathrm{T}} \Phi_{2^p}^{\mathrm{T}} H(I_{2^n} \otimes H_F) \in \mathbb{R}^{1 \times 2^{2n}}. \tag{6.49}$$

Applying the auxiliary state variable $\tilde{x}$ (6.43),

$$\tau(t) = \tilde{H} \tilde{x}(t). \tag{6.50}$$

is obtained. Based on (6.50), the set of states $x^*(t) x_F^*(t) \in \Delta_{2^{2n}}$ that lead to $Hx^*(t) \neq H_F x_F^*(t)$ is $S_x = \{\delta_{2^{2n}}^i | [\tilde{H}]_i = 1\}$.

It is worthy to mention that starting from any state in the set $\Delta_{2^{2n}} \backslash S_x$, once one of the states in the set S_x is reached, it is recognized that the fault F has occurred, no matter what the successor state is. From this point of view, the states in the set S_x are equivalent and the state space can be eliminated to reduce computational complexity, which results in a reduced augmented BCN. To this goal, one should delete the transitions between the states in the set S_x at first. For this, let the matrix P be constructed by deleting the column of the identity matrix $I_{2^{2n}}$ corresponding to the set S_x. After that, a new variable $z(t) \in \Delta_{2^{2n} - |S_x| + 1}$ is introduced which is defined as

$$z(t) = V^{\mathrm{T}}\tilde{x}(t) = \left[P\ \tilde{H}^{\mathrm{T}}\right]^{\mathrm{T}} \tilde{x}(t) \tag{6.51}$$

where $\tilde{H}$ is obtained by (6.50). It can be seen that V^{T} is a logical matrix and $z(t) = \delta_{2^{2n}-|S_x|+1}^{2^{2n}-|S_x|+1}$, $\forall\ \tilde{x}(t) = \delta_{2^{2n}}^{i} \in S_x$. Secondly, the transitions from the state in the set S_x to any state in the set $\Delta_{2^{2n}} \backslash S_x$ should be removed. Let the dynamics based on the new state space $z(t) \in \Delta_{2^{2n}-|S_x|+1}$ be governed by the following BCN without output equation

$$z(t+1) = \tilde{L}^{*}u(t)z(t) \tag{6.52}$$

where the i-th block of the matrix $\tilde{L}^{*}$, $i = 1, 2, \cdots, 2^m$ is

$$Col_j(\tilde{L}_i^{*}) = \begin{cases} Col_j(V^{\mathrm{T}}\tilde{L}_i V), & j = 1, 2, \cdots, 2^{2n} - |S_x| \\ \delta_{2^{2n}-|S_x|+1}^{2^{2n}-|S_x|+1}, & j = 2^{2n} - |S_x| + 1. \end{cases} \tag{6.53}$$

Let z_e be denoted as $\delta_{2^{2n}-|S_x|+1}^{2^{2n}-|S_x|+1}$. Therefore, if $z(t) = z_e$, then the successor state is $z(t+1) = z_e$, no matter what the input $u(t) \in \Delta_{2^m}$ is, i.e.

$$z_e = \tilde{L}^{*}u(t)z_e,\ \forall u(t) \in \Delta_{2^m}. \tag{6.54}$$

It is important to point out that the dimension of the logical matrix $\tilde{L}^{*}$ for the BCN (6.52) is $(2^{2n}-|S_x|+1)\times 2^m\cdot(2^{2n}-|S_x|+1)$, which is lower than the system matrix for the augmented BCN used in Fornasini and Valcher (2015b) (i.e. $2^{2n} \times 2^{2n+m}$).

Recalling Definition 6.16, the active fault detection problem is reformulated as finding an input sequence $\{u(0), u(1), \cdots, u(T)\}$ so that starting from any initial state, the BCN (6.52) can be stabilized at state z_e. According to Definition 2.1 (i.e. definition of STP), if the input signal $u(t)$ is selected to be $\delta_{2^m}^{i}$, then the BCN (6.52) can be expressed as

$$z(t+1) = \tilde{L}_i^{*}z(t) \tag{6.55}$$

which implies that the BCN (6.52) has a form similar to an ASLS. In the literature, the dead-beat stabilization problem of ASLSs is defined as follows.

Definition 6.18 (Fiacchini and Millérioux (2019)) Given a set of N matrices $\mathcal{A} = \{A_1, A_2, \cdots, A_N\} \subset \mathbb{R}^{n\times n}$. An ASLS described by

$$x(t+1) = A_{\sigma(t)}x(t) \tag{6.56}$$

is said to be dead-beat stabilizable if there are an integer $T \in \mathbb{N}$ and an input sequence $\{\sigma(0), \sigma(1), \cdots, \sigma(T)\}$ such that $A_{\sigma(T)} A_{\sigma(T-1)} \cdots A_{\sigma(0)} = \mathbf{0}_n \otimes \mathbf{0}_n^{\mathrm{T}}$.

Assume that the finite switching signal sequence $\sigma = \{\sigma(0), \sigma(1), \cdots, \sigma(T)\}$ can be separated into g pieces. Let γ^i represent the i-piece of the finite switching signal sequence of length h_i, i.e. $\left[[\gamma^i]_1\ [\gamma^i]_2\ \cdots\ [\gamma^i]_{h_i}\right]$, and denote $\mathbb{A}_{\gamma^i} = A_{[\gamma^i]_{h_i}} A_{[\gamma^i]_{h_i-1}} \cdots A_{[\gamma^i]_1}$. Define $\mathcal{I}_s = \{i | i \in \{1, 2, \cdots, N\}$ and $\det(A_i) = 0\}$ as the set of indices of singular matrices. In linear algebra, the image space and the null space of a matrix are defined as follows.

Definition 6.19 (Image Space (Meyer, 2000)) The image space (also known as column space) of matrix $A \in \mathbb{R}^{m \times n}$ is defined as

$$\mathrm{im}(A) = \{Ax | x \in \mathbb{R}^n\} \subseteq \mathbb{R}^m. \tag{6.57}$$

Definition 6.20 (Null Space (Meyer, 2000)) The null space (also known as kernel) of matrix $A \in \mathbb{R}^{m \times n}$ is defined as

$$\ker(A) = \{x | Ax = 0\} \subseteq \mathbb{R}^n. \tag{6.58}$$

To analyze the dead-beat stability of ASLSs, a necessary and sufficient condition is given.

Proposition 6.21 (Fiacchini and Millérioux (2019)) The system (6.56) is dead-beat stabilizable if and only if there are $g \in \mathbb{N}$ subsequences $\{\gamma^1, \gamma^2, \cdots, \gamma^g\}$ so that

$$\sum_{\alpha=1}^{g} \dim\left(\mathrm{im}\left(\prod_{i=1}^{\alpha-1} \mathbb{A}_{\gamma^i}\right) \cap \ker(\mathbb{A}_{\gamma^\alpha})\right) = n, \tag{6.59}$$

where $[\gamma^i]_{h_i}$ belongs to $\mathcal{I}_s$ for $i = 1, 2, \cdots, g$. The symbols $\mathrm{im}(\cdot)$ and $\ker(\cdot)$ denote, respectively, the image space and null space of a matrix and $\dim(\cdot)$ is the number of vectors of any basis for a vector space.

In order to apply the condition given in Proposition 6.21, the BCN (6.52) shall be slightly modified sothat it is stabilized at $\mathbf{0}_{2^{2n}-|S_x|+1}$ instead of z_e. One can obtain that

$$\mathbf{0}_{2^{2n}-|S_x|+1} = \tilde{L}^* u(t)\, \mathbf{0}_{2^{2n}-|S_x|+1}, \ \forall u(t) \in \Delta_{2^m}. \tag{6.60}$$

As $\tilde{L}^*$ is a logical matrix, the last row and column of the blocks $\tilde{L}_i^*, i = 1, 2, \cdots, 2^m$ are set to $\mathbf{0}_{2^{2n}-|S_x|+1}$. In addition, applying the modified matrix $\tilde{L}^*$, the last entry of the vector $z(t)$ calculated by (6.52) is always zero. Therefore, the last row and the last column of the modified blocks $\tilde{L}_i^*, i = 1, 2, \cdots, 2^m$ are deleted and the result is denoted as $\bar{L} \in \mathbb{R}^{(2^{2n}-|S_x|)\times 2^m\cdot(2^{2n}-|S_x|)}$. Based on the matrix $\bar{L}$, the BCN (6.52) is transformed into an ASLS described by

$$\tilde{z}(t+1) = \bar{L}u(t)\tilde{z}(t) \tag{6.61}$$

and the active fault detection problem of BCNs can be reformulated as follows.

Lemma 6.22 The active fault detection problem of BCNs is solvable if and only if there is an input sequence $\{u(0), u(1), \cdots, u(T)\}$ so that

$$\bar{L}_{u(T)}\bar{L}_{u(T-1)}\cdots\bar{L}_{u(0)} = \mathbf{0}_{2^{2n}-|S_x|} \otimes \mathbf{0}^{\mathrm{T}}_{2^{2n}-|S_x|}. \tag{6.62}$$

Let $\bar{L}_i$ be the i-th block of the matrix $\bar{L}$. $\mathcal{I}_s$ is the set of indices of singular matrices, i.e., $\mathcal{I}_s = \{i | i \in \{1, 2, \cdots, 2^m\}$ and $\det(\bar{L}_i) = 0\}$. Based on Proposition 6.21, a necessary and sufficient condition for the active fault detectability of BCNs can be given as follows.

Theorem 6.23 *The fault F is actively detectable if and only if there are $g \in \mathbb{N}$ subsequences $\{\gamma^1, \gamma^2, \cdots, \gamma^g\}$ so that*

$$\sum_{\alpha=1}^{g} \dim\left(im\left(\prod_{i=1}^{\alpha-1}\bar{\mathbb{L}}_{\gamma^i}\right) \cap \ker(\bar{\mathbb{L}}_{\gamma^\alpha})\right) = 2^{2n} - |S_x| \tag{6.63}$$

where $[\gamma^i]_{h_i}$ belongs to $\mathcal{I}_s$ for $i = 1, 2, \cdots, g$.

Remark 6.24 As a matter of fact, the BCN (2.32)–(2.33) also has a form similar to an ASLS. The stabilization problem of BCNs is to find a possible input sequence so that the BCN is convergent to a fix point x_e (see Definition 11.6 in Cheng et al. (2011)). If each block of the logical matrix L is modified by deleting the row and column corresponding to x_e, then the condition given in Theorem 6.23 can naturally be applied to check whether the state stabilization problem of the BCN (2.32)–(2.33) via a free control sequence is solvable.

6.2.2 Active Detector and Input Sequence Generator

After the necessary and sufficient condition for active detectability of BCNs is given in Section 6.2.1, an approach will be proposed here to determine an input sequence that achieves active fault detection of BCNs. First, based on the result given in Fiacchini and Millérioux (2019), an algorithm will be given to find a test input sequence $\{u(0, u(1), \cdots, u(T))\}$ whose length T is minimal if it exists. After that, an observer based approach to design an active detector will be introduced.

Recently, the dead-beat stabilization problem of autonomous switched linear discrete time systems (ASLSs) has been studied in Fiacchini and Millérioux (2019). An approach has been proposed in Fiacchini and Millérioux (2019) which requires lower computational effort than directly applying exhaustive search (for instance the brute-force approach) on the whole sequence. In the derivation, the following linear algebra matrix equality is used.

Lemma 6.25 (Meyer (2000)) Given two matrices $A \in \mathbb{R}^{m\times n}$ and $B \in \mathbb{R}^{n\times p}$, the following equality holds

$$\dim\ker(A \cdot M) = \dim(\mathrm{im}(B) \cap \ker(A)) \tag{6.64}$$

where M is a basis matrix of $\mathrm{im}(B)$.

By applying Lemma 6.25, Fiacchini and Millérioux (2019) points out that given any b finite subsequences $\{\gamma^1, \gamma^2, \cdots, \gamma^b\}$, to check

$$\dim\left(\mathrm{im}\left(\prod_{j=1}^{b-1} \mathbb{A}_{\gamma^j}\right) \cap \ker(\mathbb{A}_{\gamma^b})\right) = d_b > 0, \tag{6.65}$$

one can simply test

$$\dim\ker(\mathbb{A}_{\gamma^b} X) = d_b > 0 \tag{6.66}$$

where X is a basis of $\mathrm{im}(\prod_{j=1}^{b-1} \mathbb{A}_{\gamma^j})$. Moreover, Fiacchini and Millérioux (2019) mentioned that the subsequence γ^b satisfying (6.66) can guarantee

$$\dim\mathrm{im}\left(\prod_{j=1}^{b} \mathbb{A}_{\gamma^j}\right) < \dim\mathrm{im}\left(\prod_{j=1}^{b-1} \mathbb{A}_{\gamma^j}\right). \tag{6.67}$$

Based on the relationship between (6.65) and (6.66), the condition (6.59) can be equivalently reformulated as finding g switching signal subsequences $\gamma^i, i = 1, 2, \cdots, g$ such that

$$\sum_{\alpha=1}^{g} \dim \ker(\mathbb{A}_{\gamma^\alpha} M_{\alpha-1}) = n \tag{6.68}$$

where $M_{\alpha-1}$ is a basis matrix of $\mathrm{im}(\prod_{j=1}^{\alpha-1} \mathbb{A}_{\gamma^j})$. For simplification, Fiacchini and Millérioux (2019) has shown that one can solely focus on candidate subsequence starting and terminating with a singular matrix. In order to find g subsequences, the γ^α-th subsequence $\alpha = 1, 2, \cdots, g$ is obtained by applying exhaustive search on the subsequences instead of the whole sequence. In this way, the computational effort is reduced considerably.

Motivated by Fiacchini and Millérioux (2019) and based on the result given in Fiacchini and Millérioux (2019), an approach will be proposed to determine a sequence γ of minimal length that achieves active fault detection of BCNs. The basic idea behind the proposed approach is to find a subsequence at each step so that (6.67) is satisfied. Assume that the matrix $\bar{L}$ of the BCN (6.61) is obtained and the index set $\mathcal{I}_s$ is known. For the sake of simplicity, $\bar{\mathbb{L}}_{\gamma^i}$ with the i-th subsequence $\gamma^i = [[\gamma^i]_1 \ [\gamma^i]_2 \ \cdots \ [\gamma^i]_{h_i}]$ is denoted as $\bar{\mathbb{L}}_{\gamma^i} = \bar{L}_{[\gamma^i]_{h_i}} \bar{L}_{[\gamma^i]_{h_i-1}} \cdots \bar{L}_{[\gamma^i]_1}$. Let the set $\mathcal{Q}_0$ be initialized as $\{\bar{L}_i | i \in \mathcal{I}_s\}$. Then, one can obtain that $\dim \ker(\bar{L}_i) > 0, \forall \bar{L}_i \in \mathcal{Q}_0$. As each column of the matrix $\bar{L}$ belongs to the set $\Delta_{2^{2n}-|S_x|} \cup \{\mathbf{0}_{2^{2n}-|S_x|}\}$, the basis matrix of $\mathrm{im}(\prod_{j=1}^{i-1} \mathbb{L}_{\gamma^j})$ can be obtained simply by deleting the duplicate columns and the zero columns and will be denoted as M_{i-1}. Based on this, the subsequence γ^i with $[\gamma^i]_{h_i} \in \mathcal{I}_s$ is saved, which satisfies

$$\dim \ker(\bar{\mathbb{L}}_{\gamma^i} M_{i-1}) = d_i > 0. \tag{6.69}$$

Repeating the same procedure, if g subsequences are found so that $\sum_{j=1}^{g} d_j = 2^{2n} - |S_x|$, then this means that the matrix $\prod_{j=1}^{g} \mathbb{L}_{\gamma^j}$ is a zero matrix. Hence, active fault detection of BCNs is achieved.

For convenience, the result above is summarized in Algorithm 7.

Algorithm 7: Given the BCN (6.52) with matrices $\bar{L}_i, i = 1, 2, \cdots, 2^m$ and the index set $\mathcal{I}_s$. Determine the input sequence $\{u(0), u(1), \cdots\}$ of minimal length that achieves active fault detection of BCNs.

1. Initialize the set $\mathcal{Q}_0$ as $\{\bar{L}_i | \; i \in \mathcal{I}_s\}$ and let $t = 0$.
2. Calculate the dimension count vector $D_t \in \mathbb{R}^{|\mathcal{Q}_t| \times 1}$ indicating the number of independent vectors of a basis in the kernel $\ker(\mathcal{Q}_t)$.
3. Calculate the basis matrices M_t of the kernel of matrices in the set $\mathcal{Q}_t$.
4. Let $h_t = 1$. Check each sequence γ^{t+1} of the length h_t with $[\gamma^i]_{h_t} \in \mathcal{I}_s$ whether (6.69) holds with respect to at least one basis matrix M_t.
5. If no sequence γ^{t+1} and $h_t \leq h_{max}$ exists, then replace h_t with $h_t + 1$ and go back to step 4). If $h_t > h_{max}$, then stop. Otherwise, save the matrix $\prod_{j=1}^{t+1} \bar{\mathbb{L}}_{\gamma^j}$ satisfying (6.69) in the set $\mathcal{Q}_{[t+1]}$ and the sequence $\gamma = [\gamma^1 \; \gamma^2 \; \cdots \; \gamma^{t+1}]$. Go to step 6).
6. Calculate the dimension counter D_t. If there is at least one entry in D_t equal to $2^{2n} - |S_x|$, then return the sequence γ. Otherwise, replace t with $t + 1$ and go to step 2).

Remark 6.26 Assume that there are 2^m subsystems $\bar{L}_i, i = 1, 2, \cdots, 2^m$ and each subsequence has an average length of r (i.e. the whole sequence of length $(g-1)r$). Fiacchini and Millérioux (2019) has pointed out that Algorithm 7 requires the complexity $\mathcal{O}((g-1) \cdot |\mathcal{I}_s| \cdot 2^{rm} + |\mathcal{I}_s|)$. In comparison to that, the brute-force approach searching for the whole sequence of length $(g-1) \cdot r$ has the complexity $\mathcal{O}(2^{rm(g-1)+2m})$. Therefore, applying Algorithm 7 to solve the active fault detection problem of BCNs is computationally more efficient than the approach proposed in Fornasini and Valcher (2015b).

After a test sequence γ has been obtained, γ is considered as input sequence u for fault detection of BCNs. Observer-based fault detection technique can be applied for fault detection of BCNs as mentioned in Fornasini and Valcher (2015a). The concept of active fault detection with given a fault detector and input sequence generator can be depicted in Fig. 6.9.

Recall that the Luenberger-like observer can be described by

$$\hat{x}(t) = \Phi_{2^n}^{\mathrm{T}}(I_{2^n} \otimes H^{\mathrm{T}}) L_{eq} W_{[2^n, 2^m]} \hat{x}(t-1) u(t-1) y(t) \tag{6.70}$$

where $\hat{x}(t)$ is the estimate of the BCN state initialized as $\hat{x}(0) = \mathbf{1}_{2^n}$ and $y(t), u(t)$ are, respectively, the output and the input at time t. Based on the observer (6.70) and recalling Boolean product (2.36), an active fault detector can be designed as

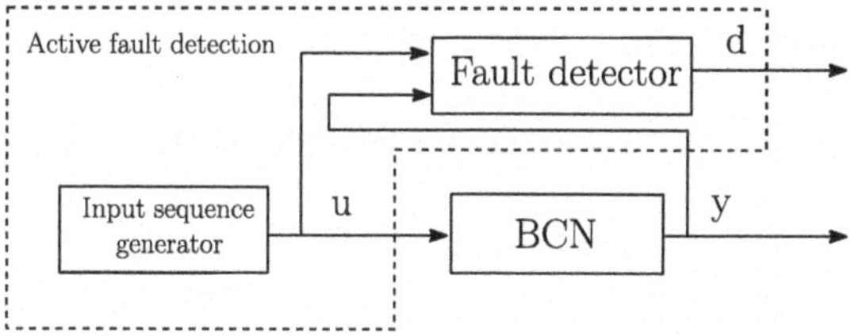

Figure 6.9 Block diagram of active fault detection with a fault detector and an input sequence generator

$$\begin{aligned} \hat{x}(t) &= \left(\Phi_{2^n}^{\mathrm{T}}(I_{2^n} \otimes H^{\mathrm{T}})L_{eq}W_{2^n,2^m}\right)\left(\hat{x}(t-1)u(t-1)y(t)\right), \\ r(t) &= 1 - \mathbf{1}_{2^n}^{\mathrm{T}}\hat{x}(t) \end{aligned} \tag{6.71}$$

where $r(t) \in \mathbb{R}$ is the residual signal.

It is important to note that if the BCN is faulty and the test sequence is applied, then the residual signal $r(t)$ converges to 1. If the BCN is fault-free and the test sequence is applied, then the residual signal $r(t)$ is less than 1. This holds independently of the initial state.

Since the upper bound of the residual signal is 1, the decision logic can be set to

$$\begin{cases} r(t) = 1 & \Rightarrow \quad \text{A fault has occurred,} \\ r(t) < 1 & \Rightarrow \quad \text{The BCN is fault-free.} \end{cases} \tag{6.72}$$

Remark 6.27 Disturbances, measurement noise and process noise often exist in practical systems. In this case, the observer proposed in Fornasini and Valcher (2015a) cannot be used any more to execute an on-line test whether a fault occurs or not. One possible idea is to consider the unknown input decoupling in observer design (i.e. unknown input observer) introduced in Section 4.2.

Remark 6.28 The active fault isolation problem of BCNs to be considered here aims at designing an input sequence so that the given k fault candidates can be isolated starting with any possible initial state. Assume that the k fault candidates are actively detectable. One possible way is to calculate the input sequence $\{\bar{u}_i(0), \bar{u}_i(1), \cdots, \bar{u}_i(T_i)\}, i = 1, 2, \cdots, k$ for the k-th fault candidate according to Algorithm 7. Based on this, an input sequence u for active fault isolation can be generated by connecting the input sequences $\bar{u}_i, i = 1, 2, \cdots, k$, i.e. $u = \{\bar{u}_1(0), \bar{u}_1(1), \cdots, \bar{u}_1(T_1), \bar{u}_2(0), \bar{u}_2(1), \cdots, \bar{u}_2(T_2), \cdots, \bar{u}_k(0), \bar{u}_k(1), \cdots, \bar{u}_k(T_k)\}$.

6.2.3 Example

In order to illustrate the main results of this section, consider the following BCN:

$$\begin{cases} X_1(t+1) = \neg X_3(t) \wedge (X_1(t) \vee U(t)), \\ X_2(t+1) = \neg X_4(t) \wedge (X_1(t) \vee X_3(t)), \\ X_3(t+1) = X_2(t), \\ X_4(t+1) = \neg X_1(t) \wedge (X_2(t) \vee X_3(t)) \end{cases} \tag{6.73}$$

which is a model for the $p53 - MDM2$ negative feedback regulatory loop in the presence of DNA double strand breaks revealed in Layek et al. (2011). Assume that the states X_1 and X_4 can be measured, while X_2 and X_3 are not measured, i.e. $Y_1(t) = X_1(t)$, $Y_2(t) = X_4(t)$.

Using the semi-tensor product of matrices, (6.73) can be converted into the form (2.34)–(2.35) with $x(t) = \ltimes_{i=1}^{3} x_i(t) \in \Delta_8$, $u(t) \in \Delta_2$, $y(t) = \ltimes_{i=1}^{2} y_i(t)$ and

$$\begin{aligned} L_{eq} &= \delta_{16}[14\ 14\ 10\ 10\ 6\ \ 6\ \ 2\ \ 2\ \ 16\ 16\ 12\ 12\ 8\ \ 8\ \ 4\ \ 4 \\ &\qquad 13\ 13\ 9\ \ 9\ 5\ 13\ 5\ 13\ 15\ 15\ 11\ 11\ 8\ 16\ 8\ 16] \\ H &= \delta_4[1\ \ 2\ 1\ \ 2\ 1\ 2\ 1\ 2\ \ 3\ \ 4\ \ 3\ \ 4\ 3\ 4\ 3\ 4]. \end{aligned} \tag{6.74}$$

Without any biological knowledge about the consequences of faults in the $p53 - MDM2$ model, assume that, as a consequence of a fault, the matrices in the faulty model (6.1)–(6.2) become

$$\begin{aligned} L_{eq} &= \delta_{16}[10\ 10\ 6\ 9\ 16\ 12\ 8\ 4\ 10\ 9\ 5\ \ 5\ 15\ 11\ 16\ \ 8 \\ &\qquad 14\ \ 10\ 6\ 2\ 16\ 12\ 8\ 4\ 13\ 9\ 13\ 13\ 15\ 11\ 16\ 16], \\ H_F &= \delta_4[1\ \ 2\ 1\ 3\ 4\ \ 2\ 1\ 2\ 3\ 4\ 3\ \ 4\ \ 1\ \ 4\ \ 3\ \ 4]. \end{aligned} \tag{6.75}$$

As first step, the reduced augmented BCN is generated. A simple calculation shows that the matrix $\bar{L}$ for the BCN (6.61) has the dimensions 64×128 which is lower than the matrix $\bar{L}$ for the augmented BCN (6.45), i.e. the dimensions 256×512.

In the next step, Algorithm 7 is executed to determine an input sequence to realize the active fault detection. One of the possible test sequences is

$$u(t) = \delta_2^1,\ u(t+1) = \delta_2^2,\ u(t+2) = \delta_2^1. \tag{6.76}$$

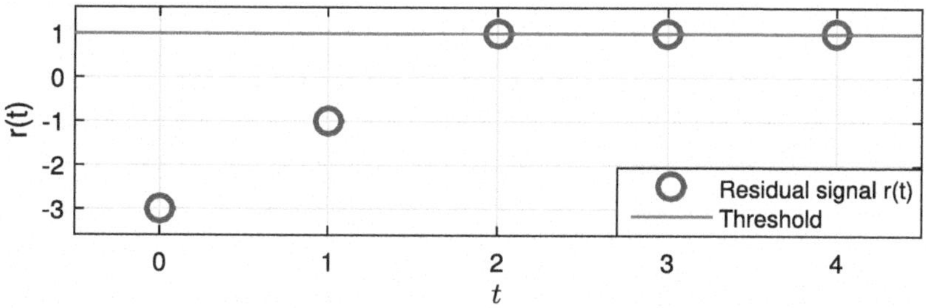

Figure 6.10 Residual signal $r(t)$ corresponding to the test sequence given in (6.76) founded by Algorithm 7

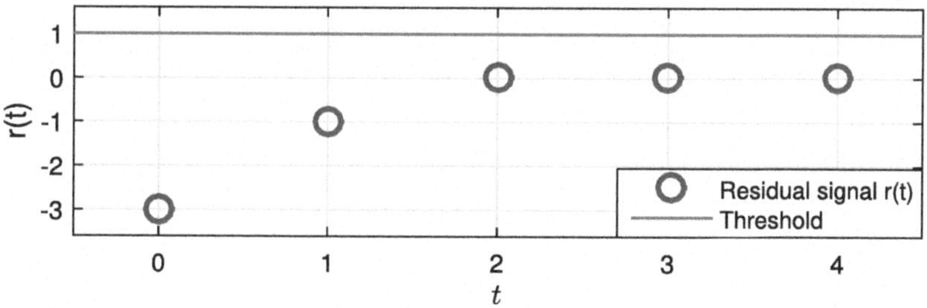

Figure 6.11 Residual signal $r(t)$ corresponding to the sequence given in (6.77)

Assume that at time $t = 0$, a fault occurs which is described by the faulty BCN (6.1)–(6.2) with the logical matrices given in (6.75). Additionally, assume that the input and the measured output at time $t = 0$, are, respectively, $u(0) = \delta_2^2$ and $y(0) = \delta_4^2$. Applying the test sequence (6.76) at time $t = 1$ (i.e. the input sequence $u(1) = \delta_2^1, u(2) = \delta_2^2, u(3) = \delta_2^1$), the corresponding residual signal is depicted in Fig. 6.10. It can be seen that applying the sequence obtained by Algorithm 7, the residual signal $r(t)$ has the value 1 at time $t = 2$. According to the decision logic (6.72), this indicates that the fault is detected at time $t = 2$. For comparison, one more sequence

$$u(1) = \delta_2^2,\ u(2) = \delta_2^2,\ u(3) = \delta_2^2 \tag{6.77}$$

is taken into consideration. The result is shown in Fig. 6.11. If the input sequence (6.77) is applied, the residual signal $r(t)$ converges to the value 0 and is below the threshold value 1. In this case, the fault cannot be detected. As a result, the test sequence (6.76) can improve the quality of fault detection.

Conclusion 7

7.1 Summary

By applying the powerful mathematical tool called semi-tensor product of matrices (STP), the dynamics of Boolean control networks (BCNs) can be converted into a model, i.e. matrix expression, that is quite similar to the standard discrete-time state space model. Hence, this tool enables researchers to systematically develop control theory for BCNs. Based on the matrix expression of BCNs, in this thesis some important control schemes of BCNs have been studied, i.e. reconstructibility analysis, observer design, output tracking control and fault diagnose.

Reconstructibility is an important system property. It is a measure of how well the current internal state of a system can be inferred from the knowledge of the input and output trajectory. At first, the relationships between the various definitions of observability and reconstructibility were comprehensively studied. The results have shown that reconstructibility and observability are not equivalent. For reconstructibility analysis, explicit and recursive methods were proposed. The recursive method shows that mapping for explicit method can be determined in an iterative procedure. Hence, it can facilitate the application to larger networks by using the recursive method. Furthermore, in order to reduce computational effort in the case of unreconstructible BCNs, stopping criteria for the recursive method were derived. It has been shown that by applying the stopping criteria the recursive method can be terminated much earlier in the case of unreconstructible BCN. For reconstructibility analysis of large-scale BCNs, one needs to partition the BCN into subnetworks of lower dimensions. Based on the subnetworks, a sufficient condition for the reconstructibility of large-scale BCNs is given.

In control theory, state observers play an important role in providing information on internal states. For observer design, an approach to design Luenberger-like

Z. Zhang, *Observer Design for Control and Fault Diagnosis of Boolean Networks*, https://doi.org/10.1007/978-3-658-35929-4_7

observers has been proposed. It was found out that the Luenberger-like observer provides an accurate state estimate no later than the minimal reconstructibility index and the real state is always included in the set of state estimates delivered by the observer. In order to reduce computational effort, an approach to design reduced-order state observers is proposed. To achieve this goal, the concept of reducible state variables was introduced. Based on this, an approach was given to find those output variables whose knowledge can be used to determine reducible state variables. After that, a method to find state and output transformation matrices was given so that BCN can be transformed into a form suitable for reduced-order observer design. Then, an approach to design reduced-order observers has been proposed. It has been shown that the state estimate provided by the reduced-order observer is equal to the state estimate obtained by the Luenberger-like observer and converges to the real state no later than the minimal reconstructibility index. The real state is always included in the state estimate provided by the reduced-order observer. The main advantage of the reduced-order observer is that the computational complexity for state estimation has been significantly reduced. For large-scale BCNs, an approach to design a distributed observer is given. Simulations have shown that the application of the distributed observer reduces memory consumption and computational effort. To consider disturbances, measurement noise and process noise, BCNs are extended to include unknown inputs. A necessary and sufficient condition for the decouplability of unknown inputs was derived. If unknown inputs are decouplable, then the unknown input observer decouples the state estimation from the effect of the unknown inputs. In addition, based on the proposed observer, it is possible to estimate the unknown inputs.

Due to the limited technique of measurement, not all the states can be directly measured. In this case, the state-feedback controller cannot be realized. However, a correct state estimation could be delivered by a state observer. Due to this the exact tracking problem and the optimal tracking problem of BCNs with time-varying reference output trajectories have been studied. Necessary and sufficient conditions for the output trajectory trackability were given. An approach to exact tracking problems have been successfully established. If the given output trajectory cannot be tracked exactly, then one can apply the optimal tracking problem formulated as ℓ_1 or ℓ_∞ optimization problems to minimize the tracking error. In addition, the optimization approaches were extended to consider changes in control inputs. Moreover, it was shown that state, transition and input constraints can be easily taken into account during the design.

In control theory, model-based fault diagnosis has been receiving more and more attention. The model-based active and passive fault diagnosis problems of BCNs were investigated. For active fault detection, a combined model was constructed by considering the non-faulty BCN and the faulty BCN together. Based on this, it was shown that the active fault detection problem of BCNs can be reformulated as a dead-beat stabilization problem of autonomous switched linear discrete-time systems. After that, a necessary and sufficient condition for active detectability of a fault was given. An algorithm to design an input sequence with minimal length for active fault detection is proposed which works more efficiently than the exhaustive search. The generated input sequence can be applied anytime to detect a given fault independent from the current state. For active fault isolation, one can connect the input sequences for each fault and, by means of the residuals, a set of faults can be isolated. The passive fault diagnosis of BCNs was investigated. By solving the relational coarsest partition (RCP) problem, one can check the observability of BCNs more efficiently. If a BCN is not observable, then the indistinguishable states can be found by solving the RCP problem. Based on this, an approach to design a reduced-order observer for passive fault detection of BCNs was proposed which requires lower computational effort than the Luenberger-like observer. After fault detection, an approach by using lower number of residual generators for passive fault isolation was given. In comparison to the conventional fault isolation scheme the main advantage of the approach proposed in the paper is that the computational complexity for fault diagnosis could be significantly reduced.

7.2 Future Work

In this thesis, a sufficient condition to check reconstructibility of large-scale BCNs with cyclic structure is proposed. However, the condition is rather conservative since it requires to cut the edges so that the modified BCN has an acyclic structure. One possible way to improve the approach for reconstructibility analysis of large-scale BCNs with cyclic structure is to analyze the convergence of the state estimation in a cycle. As reconstructibility analysis of BCNs is generally a $\mathcal{NP}$-hard problem, one challenge is to propose an efficient approach for reconstructibility analysis, especially for large-scale BCNs. One possible idea is to incorporate prior knowledge, like canalizing and unate Boolean functions. If a Boolean variable is assigned to a canalizing function, then the Boolean variable may only take canalizing output value under some circumstances.

8 Kurzfassung in deutscher Sprache (extended summary in German)

Mathematische Modellierungen biologischer Prozesse ermöglichen tiefere Einblicke zum besseren Verständnis in komplexen biologischen Systemen (z. B. zelluläre Systeme). Kontinuierliche Modelle wie z. B. die gewöhnlichen Differentialgleichungen (ODE) können zur detaillierten Beschreibung eines biologischen Prozesses angewendet werden. Allerdings sind die Abtastfrequenzen der Messdaten im Vergleich mit typischen technischen Systemen im Bereich der Elektrotechnik deutlich niedriger und die Kenntnisse der biologischen Wirkungsmechanismen fehlen. Wegen der begrenzten Menge an Messdaten sind kontinuierliche Modelle für ein biologisches System oft schwierig zu erhalten. Dadurch sind die erworbene kontinuierliche Modelle ungenau. Deshalb ist die Anwendung der modellgestützten Fehlerdiagnose generell noch schwieriger. Im Gegensatz dazu ist ein boolesches Netzwerk (BN) ein zeit-diskretes Modell. Häufig stehen nur qualitative Kenntnisse und Daten für die Untersuchung der dynamischen Interaktionen zwischen Systemkomponenten in biologischen Prozessen zur Verfügung. Dies macht das BN als eine grobe Vereinfachung der biologischen Prozessen attraktiver als kontinuierliche Modelle.

In BNs ist die Systemdynamik durch logische Verknüpfungen der booleschen Variablen (z. B. UND, ODER, NICHT, usw.) beschrieben. BNs können viele Systeme beschreiben, welche lediglich hohe Pegel und niedrige Pegel unterscheiden (z. B. zur Beschreibung von sequentiellen dynamischen Systemen). Die Zustände in solchen Systemen nehmen ausschließlich die zwei Werte „on" und „off" an, welche der Logik „wahr" und „falsch" entsprechen. Die Abhängigkeit der Dynamik des BNs von den Eingangssignalen kann mit logischen Formeln beschrieben werden. Außerdem können wegen der beschränkten Messtechnik nicht alle internen Zustandsvariablen direkt gemessen werden. Zur Berücksichtigung dieser Faktoren kann das Konzept des BNs leicht durch Einführung von Eingangssignalen und Ausgangssignalen zu einem booleschen Regelungsnetzwerk (BCN) erweitert werden. Durch

Z. Zhang, *Observer Design for Control and Fault Diagnosis of Boolean Networks*,
https://doi.org/10.1007/978-3-658-35929-4_8

die Pionierarbeit von Kaufmann (Kauffman, 1969) wurde das BN zur Beschreibung von genetischen regulativen Netzwerken im Bereich der Biologie eingeführt. Jedoch fehlt bis vor kurzem ein geeignetes und systemtechnisches Werkzeug, um das dynamische Verhalten des BNs zu analysieren und zu beobachten. In den letzten zehn Jahren wurde das mathematische Tool "das Semi-Tensorprodukt von Matrizen" (STP) eingeführt (Cheng et al., 2012). Durch STP können BCNs in einer Form ähnlich wie ein Zustandsraummodell dargestellt werden, welche es ermöglicht, die grundlegenden Systemeigenschaften der BCNs (z. B. Beobachtbarkeit, Rekonstruierbarkeit usw.) zu analysieren und die regelungstechnischen Methoden (z. B. Beobachter- und Reglerentwurf) auf BCNs anzuwenden.

Rekonstruierbarkeit ist eine Systemeigenschaft, welche beschreibt, ob Zustände in ein System aus der Kenntnis der messbaren Größen im Zeitverlauf rekonstruiert werden können. In der Literatur sind verschiedene Definitionen der Rekonstruierbarkeit und Beobachtbarkeit veröffentlicht. Die Beziehung zwischen der Beobachterbarkit in Definition 3.1-Definition 3.4 und der Rekonstruierbarkeit in Definition 3.5-Definition 3.7 ist in Abbildung 8.1 gezeigt. Im Allgemeinen führt Beobachtbarkeit zu Rekonstruierbarkeit. Zur Analyse der Performanz des zu entwerfenden Beobachters für ein BCN ist die Erfüllung der Rekonstruierbarkeit erforderlich. Daher in dieser Arbeit ist eine effiziente Methode zur Analyse der Rekonstruierbarkeit vorgestellt. Die Grundidee liegt darin, dass sich die Beziehung zwischen

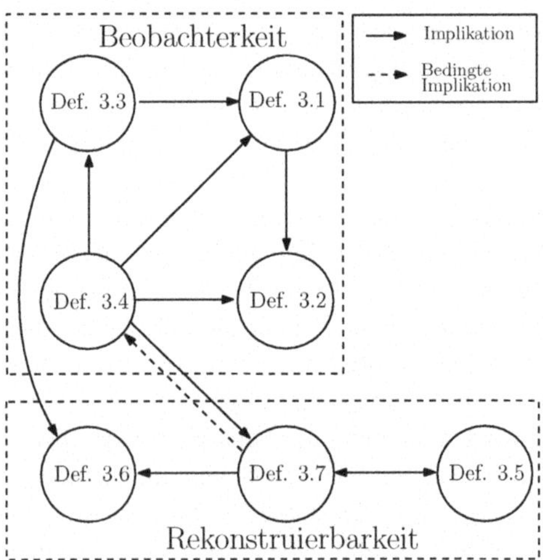

Figure 8.1 Die Beziehungen zwischen Beobachterbarkit und Rekonstruierbarkeit

Endzuständen und den Eingangs- und Ausgangstrajektorien durch einen iterativen Prozess berechnen lassen kann. Zusätzlich wird ein Abbruchkriterium eingeführt. Für ein nicht-rekonstruierbares BCN bringt das Kriterium einen Vorteil, dass der iterative Prozess frühzeitig beendet wird, womit der Rechenaufwand reduziert werden kann.

Ein Zustandsbeobachter ist ein System, das aus bekannten Eingangsgrößen und Ausgangsgrößen eines beobachteten Referenzsystems die internen Zustände rekonstruiert. Ein Zustandsbeobachter für BCNs kann nur entworfen werden, wenn BCNs rekonstruierbar sind. In dieser Arbeit ist ein Ansatz zum Entwurf von Luenberger-like Beobachter entwickelt. Der Prinzip hinter dem Luenberger-like Beobachter basiert sich auf den zwei Schritten (Prädiktion und Korrektur). Ohne die Information über Eingangsgrößen und Ausgangsgrößen lässt sich der Zustand des Luenberger-like Beobachters mit einer Menge aller möglichen Zustände initialisieren. Zwecks der Prädiktion ist eine Vorhersage der möglichen Zustände für den aktuellen Zeitpunkt anhand des mathematischen Modells für BCN enthalten. Danach werden zur Korrektur die Zustände ausgeschlossen, die nicht mit den neuen Informationen der aktuell gemessenen Eingangsgrößen und Ausgangsgrößen zusammenpassen. Für ein rekonstruierbares BCN, liefert der Luenberger-like Beobachter die korrekte Zustandsschätzung. Dabei kennzeichnet der Rekonstruierbarkeitsindex die Konvergenzgeschwindigkeit der zu liefernden Zustandsschätzung konvergiert. Je größer das System ist, desto stärker wächst der Rechenaufwand der entworfenen Beobachter. Für bessere Umsetzung der Online Zustandsschätzung, ist es vorteilhaft, den Rechenaufwand zu reduzieren. Zu diesem Zweck wurde ein Teil der Zustandsvariablen (reduzierbare Zustandsvariablen) des BCNs identifiziert, welche direkt gemessen werden können bzw. sich durch Ausgangssignale unmittelbar berechnen lassen. Darauf basierend wird ein Beobachter mit reduzierter Ordnung entwickelt. Die Eigenschaften des Beobachters mit reduzierter Ordnung sind identisch mit denen des vollständigen Luenberger-like Beobachters (z. B. Konvergenzgeschwindigkeit). Allerdings reduziert sich der entsprechende Rechenaufwand exponentiell in Bezug auf die Anzahl der reduzierbaren Zustandsvariablen. In der realen Welt ist ein biologisches System so groß, dass der Luenberger-like Beobachter nicht direkt anwendbar ist. Um dieses Problem zu lösen, kann ein verteilter Beobachter entworfen werden. Davor muss das gesamte System in verschiedene kleine Teilsysteme zerlegt werden. Für jedes Teilsystem wird ein Luenberger-like Beobachter entworfen, welche mit Nachbarteilsystemen über die Zustandsschätzung kommuniziert. Im Vergleich zum Luenberger-like Beobachters hat der verteilte Beobachter einen deutlich reduzierten Rechenaufwand, besonders wenn für jedes Teilsystem ein Beobachter mit reduzierter Ordnung verwendet wird. Allerdings ist nach der Zerlegung des gesamten Systems der Rekonstruierbarkeitsindex schwieriger einzuschätzen. Im Fall eines auftretenden Rauschens kann der Luenberger-like Beobachter leicht zu einen Unknown Input Observer (UIO) erweitert werden, welche alle Effekte des Rauschens berücksichtigt werden.

In vielen Fälle können wegen der beschränkten Messtechnik nicht alle Zustände direkt gemessen werden. In diesem Fall kann ein Zustandsregler nicht realisiert werden. Um dies zu ermöglichen, kann man die Zustände mit Hilfe eines Luenberger-like Beobachters rekonstruieren. In dieser Arbeit wird vorausgesetzt, dass die korrekte Zustandsschätzung verfügbar ist. Ein auf STP basierender Ansatz zum Entwurf eines Zustandsreglers für BCNs wird vorgestellt, um eine zeitvariante Ausgangstrajektorie mit endlicher Länge zu verfolgen. Zunächst werden die Bedingungen für die Erreichbarkeit der vorgegebenen Ausgangstrajektorie vorgestellt. Falls die Ausgangstrajektorie erreichbar ist, wird die Methode zur exakten Folgeregelung angewendet. Anderenfalls wird ein ℓ_1 oder ℓ_∞ Optimierungsproblem aufgestellt. Die ℓ_1 und ℓ_∞ Optimierungsprobleme können durch dynamische Programmierung effizient gelöst werden. Die optimale Lösung für das ℓ_1 Optimierungsproblem führt zu Minimierung des gesamten Verfolgungsfehlers. Im Gegensatz dazu kann durch das ℓ_∞ Optimierungsproblem der gewünschte Regler so hergeleitet werden, dass der maximale Verfolgungsfehler minimiert wird. Außerdem bringen die zwei Methoden den Vorteil, dass Beschränkungen an Eingangssignalen und Zuständen einfach berücksichtigt werden können.

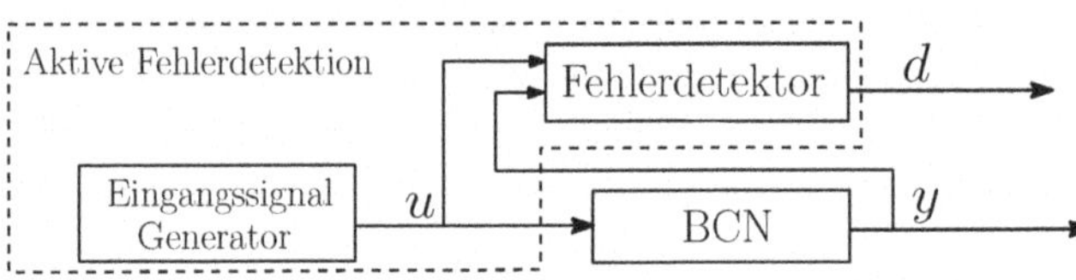

Figure 8.2 Blockdiagramm der aktiven Fehlerdetektion mit einem Fehlerdetektor und einem Eingangssignal Generator

Ein Systemfehler verursacht ein unerlaubtes Fehlerverhalten (Abweichung vom normalen Verhalten). Mittels Fehlerdiagnose lassen sich Fehler erkennen und isolieren. Modellbasierte Fehlerdiagnose ist ein in der Forschung weit angewendetes Konzept zur Fehlererkennung und Fehlerisolation. Zum einen soll das zu entwerfende Fehlerdiagnosesystem das Fehlerverhalten eines Systems automatisch erkennen. Zum anderen soll der Fehler anhand des Fehlerverhaltens diagnostiziert werden. In dieser Arbeit werden Ansätze zur aktiven und passiven Fehlerdiagnose in BCNs durch Anwendung des STPs vorgestellt. Das Ziel der aktive Fehlerdiagnose ist zum Entwurf einer Eingangssequenz, so dass Fehler schnell bemerkt werden kann. Zu diesem Zweck wird das Konzept der aktiven Fehlerdetektierbarkeit zuerst eingeführt. Danach wird das Dead-Beat Stabilisierungsproblem von autonomen geschalteten linearen zeitdiskreten Systemen vorgestellt. Durch mathematische Herleitung wird gezeigt, dass das Problem zur aktiven Fehlerdetektion von BCNs zu einem Dead-Beat Stabilisierungsproblem von autonomen geschalteten linearen zeitdiskreten Systemen umformuliert werden kann. Darauf basierend sind notwendige und hinreichende Bedingungen für die aktive Fehlerdetektierbarkeit eines Fehlers gegeben. Ein System zur aktiven Fehlerdetektion ist wie in Abbildung 8.2 aufgebaut. Dabei wird der Entwurf eines Generators der Eingangssequenz und aktiven Fehlerdetektors vorgestellt. Der aktive Fehlerdetektor verwendet die aktuelle Zustandsschätzung und konstruiert Residuen für die Fehlerdetektion. Dies bringt den Vorteil, dass die Eingangssequenz zur aktiven Fehlerdetektion unabhängig vom aktuellen Systemzustand einsetzbar ist. Zur aktiven Fehlerisolation werden Eingangssequenzen für alle Fehlerkandidaten berechnet und nacheinander angewendet. Im Vergleich dazu wird bei der passiven Fehlerdiagnose kein Eingangsignal ausgerechnet. Hierfür wurde eine Methode entwickelt um den benötigten Rechenaufwand zu reduzieren. Die Systemstruktur für die passive Fehlerdetektion ist in Abbildung 8.3 gezeigt. Zunächst wird das Relational Coarsest Partition (RCP) Problem von Transitionssystemen zuerst kurz vorgestellt. Es wird hergeleitet, dass identische Systemdynamiken innerhalb eines BCNs durch Lösung eines RCP Problems festgestellt werden können. Basierend darauf wird ein Ansatz zu Entwurf eines Fehlerdetektors entwickelt. Der Rechenaufwand für passive Fehlerdetektion kann dadurch reduziert werden, dass die identische Systemdynamik zwischen fehlerhaftem und normalem System vernachlässigt wird. Für die passive Fehlerisolation wird Entwurf der Fehlerisolationsbeobachter vorgestellt. Bei den Fehlerkandidaten existieren häufig identische Systemdynamiken. Nach der Lösung der RCP Problems werden die Dynamiken der Fehlerkandidaten in einige nebenläufigen Teilsystemen aufgeteilt bzw. gruppiert. Dadurch kann der Rechenaufwand erheblich reduziert werden.

Figure 8.3 Blockdiagramm der passiven Fehlerdetektion mit einem Fehlerdetektor

Bibliography

Abdollahi, J. and Dubljevic, S. (2012). "Lipid production optimization and optimal control of heterotrophic microalgae fed-batch bioreactor". In: *Chemical Engineering Science* vol. 84, pp. 619–627.

Akutsu, T., Hayashida, M., Ching, W.-K., and Ng, M. K. (2007). "Control of Boolean networks: hardness results and algorithms for tree structured networks". In: *Journal of Theoretical Biology* vol. 244, no. 4, pp. 670–679.

Bornholdt, S. (2008). "Boolean network models of cellular regulation: prospects and limitations". In: *Journal of the Royal Society Interface* vol. 5, no. suppl_1, S85–S94.

Chen, H., Li, X., and Sun, J. (2015). "Stabilization, Controllability and Optimal Control of Boolean Networks With Impulsive Effects and State Constraints". In: *IEEE Transactions on Automatic Control* vol. 60, no. 3, pp. 806–811.

Chen, W.-H., Yang, J., Guo, L., and Li, S. (2016). "Disturbance observer-based control and related methods–An overview". In: *IEEE Transactions on Industrial Electronics* vol. 63, no. 2, pp. 1083–1095.

Cheng, D. and Qi, H. (2009). "Controllability and observability of Boolean control networks". In: *Automatica* vol. 45, no. 7, pp. 1659–1667.

Cheng, D. and Qi, H. (2010). "A Linear Representation of Dynamics of Boolean Networks". In: *IEEE Transactions on Automatic Control* vol. 55, no. 10, pp. 2251–2258.

Cheng, D., Qi, H., and Li, Z. (2011). *Analysis and control of Boolean networks: A semi-tensor product approach*. London: Springer.

Cheng, D., Qi, H., and Zhao, Y. (2012). *An introduction to semi-tensor product of matrices and its applications*. Singapore: World Scientific.

Cheng, D., Zhao, Y., and Xu, T. (2015). "Receding Horizon Based Feedback Optimization for Mix-valued Logical Networks". In: *IEEE Transactions on Automatic Control* vol. 60, no. 12, pp. 3362–3366.

Cheng, D., Qi, H., Liu, T., and Wang, Y. (2016). "A note on observability of Boolean control networks". In: *Systems & Control Letters* vol. 87, pp. 76–82.

Falcao, D. M., Wu, F. F., and Murphy, L. (1995). "Parallel and distributed state estimation". In: *IEEE Transactions on Power Systems* vol. 10, no. 2, pp. 724–730.

Faryabi, B., Vahedi, G., Chamberland, J.-F., Datta, A., and Dougherty, E. R. (2008). "Optimal constrained stationary intervention in gene regulatory networks". In: *EURASIP journal on bioinformatics & systems biology*, p. 620767.

Z. Zhang, *Observer Design for Control and Fault Diagnosis of Boolean Networks*,
https://doi.org/10.1007/978-3-658-35929-4

Fernandez, J.-C. (1990). "An implementation of an efficient algorithm for bisimulation equivalence". In: *Science of Computer Programming* vol. 13, no. 2, pp. 219–236.

Fiacchini, M. and Millérioux, G. (2019). "Dead-Beat Stabilizability of Discrete-Time Switched Linear Systems: Algorithms and Applications". In: *IEEE Transactions on Automatic Control* vol. 64, no. 9, pp. 3839–3845.

Fornasini, E. and Valcher, M. E. (2013). "Observability, Reconstructibility and State Observers of Boolean Control Networks". In: *IEEE Transactions on Automatic Control* vol. 58, no. 6, pp. 1390–1401.

Fornasini, E. and Valcher, M. E. (2014a). "Feedback stabilization, regulation and optimal control of Boolean control networks". In: *Proc. of the 2014 American Control Conference*, pp. 1981–1986.

Fornasini, E. and Valcher, M. E. (2014b). "Optimal Control of Boolean Control Networks". In: *IEEE Transactions on Automatic Control* vol. 59, no. 5, pp. 1258–1270.

Fornasini, E. and Valcher, M. E. (2015a). "Fault Detection Analysis of Boolean Control Networks". In: *IEEE Transactions on Automatic Control* vol. 60, no. 10, pp. 2734–2739.

Fornasini, E. and Valcher, M. E. (2015b). "Fault detection problems for Boolean networks and Boolean control networks". In: *Proc. of the 34th Chinese Control Conference*, pp. 1–8.

Franke, D. (1995). "A linear state space approach to a class of discrete-event systems". In: *Mathematics and computers in simulation* vol. 39, no. 5–6, pp. 499–503.

Hamming, R. W. (1950). "Error detecting and error correcting codes". In: *Bell Labs Technical Journal* vol. 29, no. 2, pp. 147–160.

Kauffman, S. A. (1969). "Metabolic stability and epigenesis in randomly constructed genetic nets". In: *Journal of Theoretical Biology* vol. 22, no. 3, pp. 437–467.

Laschov, D., Margaliot, M., and Even, G. (2013). "Observability of Boolean networks: A graph-theoretic approach". In: *Automatica* vol. 49, no. 8, pp. 2351–2362.

Layek, R., Datta, A., and Dougherty, E. R. (2011). "From biological pathways to regulatory networks". In: *Molecular BioSystems* vol. 7, no. 3, pp. 843–851.

Li, H., Wang, Y., and Xie, L. (2015). "Output tracking control of Boolean control networks via state feedback: Constant reference signal case". In: *Automatica* vol. 59, pp. 54–59.

Li, R., Yang, M., and Chu, T. (2013). "State Feedback Stabilization for Boolean Control Networks". In: *IEEE Transactions on Automatic Control* vol. 58, no. 7, pp. 1853–1857.

Liang, J., Chen, H., and Liu, Y. (2017). "On algorithms for state feedback stabilization of Boolean control networks". In: *Automatica* vol. 84, pp. 10–16.

Luenberger, D. (1971). "An introduction to observers". In: *IEEE Transactions on Automatic Control* vol. 16, no. 6, pp. 596–602.

Meyer, C. D. (2000). *Matrix analysis and applied linear algebra*. Vol. 71. Siam.

Pahl, P. J. and Damrath, R. (2001). *Mathematical foundations of computational engineering: a handbook*. New York: Springer.

Paige, R. and Tarjan, R. E. (1987). "Three partition refinement algorithms". In: *SIAM Journal on Computing* vol. 16, no. 6, pp. 973–989.

Sridharan, S., Layek, R., Datta, A., and Venkatraj, J. (2012). "Boolean modeling and fault diagnosis in oxidative stress response". In: *BMC genomics* vol. 13 Suppl 6, S4.

Tarjan, R. E. (1972). "Depth-First Search and Linear Graph Algorithms". In: *SIAM Journal on Computing* vol. 1, no. 2, pp. 146–160.

Toutain, P.-L. and BOUSQUET-MÉLOU, A. (2004). "Plasma terminal half-life". In: *Journal of veterinary pharmacology and therapeutics* vol. 27, no. 6, pp. 427–439.

Veliz-Cuba, A. and Stigler, B. (2011). "Boolean models can explain bistability in the lac operon". In: *Journal of computational biology* vol. 18, no. 6, pp. 783–794.

Wang, Y., Liu, T., and Cheng, D. (2016). "Some notes on Semi-tensor product of matrices and SWAP matrix". In: *J. Sys. Sci. & Math. Scis. (in Chinese)* vol. 36, o. 9, 1367, pp. 1367–1375.

Weiss, E. and Margaliot, M. (2019). "Output Selection and Observer Design for Boolean Control Networks: A Sub-Optimal Polynomial-Complexity Algorithm". In: *IEEE Control Systems Letters* vol. 3, no. 1, pp. 210–215.

Wynn, M. L., Consul, N., Merajver, S. D., and Schnell, S. (2012). "Logic-based models in systems biology: a predictive and parameter-free network analysis method". In: *Integrative Biology* vol. 4 (11), pp. 1323–1337.

Xiao, Y. and Dougherty, E. R. (2007). "The impact of function perturbations in Boolean networks". In: *Bioinformatics* vol. 23, no. 10, pp. 1265–1273.

Xu, X. and Hong, Y. (2013). "Observability analysis and observer design for finite automata via matrix approach". In: *IET Control Theory & Applications* vol. 7, no. 12, pp. 1609–1615.

Zhang, K. and Zhang, L. (2016). "Observability of Boolean Control Networks: A Unified Approach Based on Finite Automata". In: *IEEE Transactions on Automatic Control* vol. 61, no. 9, pp. 2733–2738.

Zhang, K., Zhang, L., and Su, R. (2016). "A Weighted Pair Graph Representation for Reconstructibility of Boolean Control Networks". In: *SIAM Journal on Control and Optimization* vol. 54, no. 6, pp. 3040–3060.

Zhao, Y., Ghosh, B. K., and Cheng, D. (2016). "Control of Large-Scale Boolean Networks via Network Aggregation". In: *IEEE Transactions on Neural Networks and Learning Systems* vol. 27, no. 7, pp. 1527–1536.

Zhao, Y., Qi, H., and Cheng, D. (2010). "Input-state incidence matrix of Boolean control networks and its applications". In: *Systems & Control Letters* vol. 59, no. 12, pp. 767–774.

Zhu, W. and Jiang, Z.-P. (2015). "Event-Based Leader-following Consensus of Multi-Agent Systems with Input Time Delay". In: *IEEE Transactions on Automatic Control* vol. 60, no. 5, pp. 1362–1367.

Zorzi, M. and Sepulchre, R. (2016). "AR Identification of Latent-Variable Graphical Models". In: *IEEE Transactions on Automatic Control* vol. 61, no. 9, pp. 2327–2340.

The manufacturer's authorised representative in the EU is Springer Nature Customer Service Centre GmbH, Europaplatz 3, 69115 Heidelberg, Germany. If you have any concerns regarding our products, please contact ProductSafety@springernature.com

Printed and bound by CPI Group (UK) Ltd, Croydon, CR0 4YY
13/08/2026
02195499-0002